Running Buildings on Natural Energy

New thinking is essential if we are to design and occupy buildings that can keep us safe with unpredictable economies, climates, energy systems and resource challenges. For too long designers have relied on mechanical solutions for heating, cooling and ventilating buildings. The twenty-first-century dream has to be of a better architecture that enables buildings to be run for as much of a day or year as possible on local, clean, reliable and affordable natural energy. Examples are included from different climates where the fundamental building design is right, its orientation, opening sizes, mass and its natural ventilation systems and pathways. Many modern buildings are poorly designed for climate as manifested by growing incidences of overheating experienced indoors, explored here. The inability of many rating systems to record and improve the climatic design of buildings raises questions about how they deal with issues of basic building performance. This book points the way towards how we can understand such problems, and move forward from over-mechanised poorly designed buildings to a new generation of adaptable buildings designed and refurbished to run largely on natural energy and capable of evolving over time to keep their occupants safe and comfortable, even in a warming world.

The chapters in this book were originally published in *Architectural Science Review*.

Sue Roaf (B.A.Hons, A.A. Dipl., PhD, ARB, FRIAS) is Emeritus Professor at Heriot Watt University, UK, and sits on the Architects Registration Board. An award-winning architect, teacher, author and activist, she has written and edited 20 books on ice-houses, energy efficiency, ecohouse, solar and sustainable design, thermal comfort and climate change adaptation.

Fergus Nicol is a Professor at London Metropolitan and Oxford Brookes University, UK. He is internationally known for his 'Adaptive' thermal comfort research which has informed National, European and International comfort standards. He teaches, and publishes, widely and his current work on overheating is internationally influential. He convenes the NCEUB network (www.nceub.org) on comfort.

Running Buildings on Natural Energy

Design Thinking for a Different Future

Edited by
Sue Roaf and Fergus Nicol

Routledge
Taylor & Francis Group

LONDON AND NEW YORK

First published 2018 by Routledge

2 Park Square, Milton Park, Abingdon, Oxon OX14 4RN
605 Third Avenue, New York, NY 10017

Routledge is an imprint of the Taylor & Francis Group, an informa business

First issued in paperback 2021

Publisher's Note

The publisher has gone to great lengths to ensure the quality of this reprint but points out that some imperfections in the original copies may be apparent.

British Library Cataloguing in Publication Data
A catalogue record for this book is available from the British Library

ISBN 13: 978-0-8153-9603-1 (hbk)
ISBN 13: 978-0-367-53011-2 (pbk)

Typeset in Myriad Pro
by diacriTech, Chennai

Publisher's Note
The publisher accepts responsibility for any inconsistencies that may have arisen during the conversion of this book from journal articles to book chapters, namely the possible inclusion of journal terminology.

Disclaimer
Every effort has been made to contact copyright holders for their permission to reprint material in this book. The publishers would be grateful to hear from any copyright holder who is not here acknowledged and will undertake to rectify any errors or omissions in future editions of this book.

Contents

Citation Information

The chapters in this book were originally published in *Architectural Science Review*, volume 60, issue 3 (2017). When citing this material, please use the original page numbering for each article, as follows:

Chapter 9

Saving energy with a better indoor environment
Gary J. Raw, Clare Littleford and Liz Clery
Architectural Science Review, volume 60, issue 3 (April 2017) pp. 239–248

Chapter 10

Ventilation strategies for a warming world
Richard Aynsley and John J. Shiel
Architectural Science Review, volume 60, issue 3 (April 2017) pp. 249–254

For any permission-related enquiries please visit:
http://www.tandfonline.com/page/help/permissions

Notes on Contributors

E. E. Alders acquired her PhD at the Delft University of Technology, the Netherlands. Currently she runs her own company for education, design and consultancy for energy and building physics.

Richard Aynsley is a retired architect and consultant with over 40 years of international experience and has authored numerous publications. He has held the position of UNESCOP Professor of Tropical Architecture, served on the US ASHRAE Standard 55 Thermal Comfort Committee and is a life member of ASHRAE and an associate member of ASCE.

Nehaa Bhavaraju completed her diploma in Architecture from the National Institute of Technology, India, and completed her masters in Advanced Sustainable Design from the University of Edinburgh, Scotland. She is a LEED Green Associate and works as an architect at Studio Naqshbandi.

Luisa Brotas read architecture in Portugal and has a PhD in Daylighting and Planning by the London Metropolitan University, UK. Luisa is Co-Director (with Prof Fergus Nicol) of the Low Energy Architecture Research Unit, LEARN, at London Metropolitan University, UK, where she lectures PhD students.

Liz Clery is a Research Director at NatCen Social Research, UK, and a Co-Director of the British Social Attitudes survey series.

Mona Doctor-Pingel completed her undergraduate studies in Architecture at the Centre for Environment and Planning (CEPT), Ahmedabad, and post-graduation in Appropriate Technology from Flensburg University, Germany.

Jarek Kurnitski is a full professor at Tallinn University of Technology, Estonia, and adjunct professor at Aalto University, Finland. He is the leader of the Nearly Zero Energy Buildings (nZEB) research group, which today operates at both universities. He is the leader of Estonian Center of Excellence in Research ZEBE (Zero Energy and Resource Efficient Smart Buildings and Districts) operating 2016–2022.

Stanley R. Kurvers is a researcher at the Faculty of Architecture, Delft University of Technology, the Netherlands, and Chair of Indoor Environment, where he contributed to the development of tools, guidelines and methods for building indoor environmental quality.

Kalle Kuusk is a senior researcher at the Department of Civil Engineering and Architecture, Tallinn University of Technology, Estonia. His research interests are building's service systems, indoor climate and energy performance of buildings.

Hugo Lavocat works at Atelier Ten in Singapore as an environmental designer to continue developing its know-how in designing low-energy buildings in tropical areas and gain international exposure.

Clare Littleford is a Senior Researcher in Longitudinal Studies at NatCen Social Research, UK. Previously she worked as a Research Associate in Human Factors in Smart Energy Systems at the UCL Energy Institute, where she worked on the project reported here.

Behdad Moghtaderi is the Head of Chemical Engineering, the Director of the Priority Research Centre for Frontier Energy Technologies & Utilisation and the Director of VTara Clean Energy Technology Centre at the University of Newcastle, Australia. He collaborates with industry and government in energy and environment, and has 5 patents and published over 300 articles.

Fergus Nicol is a Professor at London Metropolitan and Oxford Brookes University, UK. He is internationally known for his 'Adaptive' thermal comfort research which has informed National, European and International comfort standards. He teaches, and publishes, widely and his current work on overheating is internationally influential. He convenes the NCEUB network (www.nceub.org) on comfort.

Adrian Page is an internationally recognized researcher in the field of structural masonry at the University of Newcastle, Australia. Over the past 35 years, he has produced over 200 publications in areas related to the fundamental and applied aspects of masonry including thermal performance. He has close links with the masonry industry and Chairs the AS3700 Masonry Structures Code committee.

Gary J. Raw is a chartered psychologist and associate fellow of the British Psychological Society, UK. He is a Visiting Professor at the UCL Energy Institute, UK, and an independent research consultant, specialising in environmental psychology.

Sue Roaf (B.A. Hons, A.A. Dipl., PhD, ARB, FRIAS) is Emeritus Professor at Heriot Watt University, UK, and sits on the Architects Registration Board. An award-winning architect, teacher, author and activist she has written and edited 20 books on ice-houses, energy efficiency, ecohouse, solar and sustainable design, thermal comfort and climate change adaptation.

Paola Sassi teaches and undertakes research at Oxford Brookes University, the Oxford Institute of Sustainable Development, and the Centre for Alternative Technology in Machynlleth, Wales.

John J. Shiel is the Principal of EnviroSustain which consults on low-carbon buildings and precincts, and was formerly an engineer and IT consultant. He is a PhD candidate at the University of Newcastle, Australia and a member of iiSBE.

Raimo Simson is a PhD candidate and researcher at the Department of Civil Engineering and Architecture, Tallinn University of Technology, Estonia. His research interests are indoor climate, energy performance of buildings and HVAC systems.

Peter J. W. van den Engel is a consultant/air flow specialist at Deerns consulting engineers and teacher/researcher at Delft University of Technology, the Netherlands.

Running buildings on natural energy: design thinking for a different future

Sue Roaf and Fergus Nicol

Twenty-five years ago in 1992 the Thermal Comfort Unit at Oxford Brookes was born from a simple desire to change the world, to intervene in the seemingly mad drift towards the universal air-conditioning of commercial buildings and to help shift the designers back towards a saner world in which buildings were run for as much of the year as possible using clean, free, sustainable, renewable energy, back to a world of largely naturally ventilated buildings. Sue Roaf, Fergus Nicol and Michael Humphreys worked with a team who over the years systematically not only provided the theoretical underpinnings for such a move through their work on thermal comfort in different climates but also sought to show through demonstrations and publications that way forward (Humphreys, Nicol, and Roaf 2015). Back then there was palpable hostility encountered when giving papers promoting the importance of using natural energy to power buildings (e.g. Roaf 1993). We naively did not recognize that this simple idea threatened the profitability of the Heating, Ventilation and Air-Conditioning (HVAC) engineer's business model: Services engineers are still paid largely according to how much mechanical plant and infrastructure they put into a building. It might appear now that we did badly in the battle of the opposing building conditioning approaches as 25 years later even houses are now forced by rating systems, regulations and fashion into using machines to ventilate them, even in temperate climates on green field sites. Extraordinary. Why would anyone fix shut the windows in their homes and pay to run a machine that gives them worse air quality than if they simply opened the window? This at a time when the spending power of our incomes is reducing for most of us and energy bills continue to soar almost regardless of market trading prices for energy? Where are the architects in this debate? How could they have let this happen? Where is the Architectural Sense and Science in it all?

The transfer of duties between architects and engineers began long ago in the mid-nineteenth century America, in a process well described by Fitch in his wonderful books on the history of American architecture (Fitch 1966, 1977). By the twentieth century the gradual but whole-scale absorption of the responsibility for the environmental performance of buildings by the HVAC engineer appeared complete to the extent that some architecture schools today do not even teach basic building science, concentrating on history, theory, philosophy, graphic design and computer modelling. The power grab of the role of the environmental designer by service engineers from the grip of the architect has been well described by many authors and was outlined in our *ASR* paper on twentieth-century standards for thermal comfort in 2010 (Roaf et al. 2010) and in our book on *Adapting Buildings and Cities for Climate Change* under the banner of 'The Emperor has got no Clothes' (Roaf, Crichton, and Nicol 2009).

The regulatory imperative to replace opening windows with fixed 'glazing' and provide year round ventilation with an HVAC plant was to a large extent driven by the thermal comfort standards that were developed by the air-conditioning industry to enable engineers to determine the best temperatures at which to set building thermostats. That research resulted in regulations based on the idea that 'people' could only be considered comfortable if room temperatures were kept within a narrow thermal band, ranging typically around 20–24°C. The assumptions underlying this research, too often peddled as irrefutable scientific fact, were based on the outputs of laboratory experiments and steady-state calculations using a limited range of physical measures relating to the space occupied and assumed clothing, occupation and activity levels. These factors were chosen and developed to balance the necessary heat flows into and out of the body to achieve comfort in a given set of climatic conditions. This is the Heat Balance approach. Completely ignored by it are the behavioural, attitudinal or psycho-physiological aspects of human interactions with each other and the buildings and the tendency of people to manage their environments to maintain their own personal comfort. From the 1960s this method, largely coming under the banner of the Predicted Mean Vote (PMV) method reigned supreme and designers were required to ensure that such temperatures could be achieved year round in all climates. The more extreme the climate the greater the HVAC plant needed to achieve these temperatures in buildings that are increasingly 'modern', light weight, over-glazed and energy hungry. As the plant got more expensive, the buildings got cheaper and one of the first things to go were the opening windows as they were deemed an unnecessary expense and the developer wanted to maximize lettable space and a lowest first cost regardless of the downstream impacts on the running costs, indoor air quality and quality of life of the building occupants. By the turn of the twenty-first century the average American office worker took one day a month off to deal with sick building syndrome symptoms and health risks that proliferated in US office buildings (Mendell et al. 2008; Mendell and Mirer 2008).

The alternative and competing theory to the Steady State or Heat Balance approach is that of the Adaptive Thermal Comfort method. Conclusions on whether people were comfortable or not in this method are based on the results of field studies during which physical parameters are measured while people are simultaneously asked whether they are comfortable or not during the course of their everyday lives at home or at work. The range of occupied comfort temperatures that result from this approach is far wider with people recorded as registering being comfortable at a huge range of indoor temperatures across climates ranging from the mid-thirties down to the mid-teens Celsius. The driver here is that people tend to occupy, in their ordinary lives, temperatures they find comfortable, and if they are not comfortable then they either change themselves, their clothes, activities or location or they change the spaces/buildings they occupy to enable them to be comfortable again if possible. People adapt to the temperatures within buildings and they adapt the buildings to suit their own thermal preferences. For more detail on these issues see the definitive introduction to Adaptive Thermal Comfort by Nicol, Humphreys, and Roaf (2012).

The Heat Balance method discourages natural ventilation while the Adaptive method enables it and there are International Comfort Standards for both. Standard ISO 7730 (BSI 2005) for the Heat Balance model and ASHRAE/ANSI standard 55 (ASHRAE 2010) and CEN Standard 15251 (BSI 2007) both include versions of the Adaptive approach. Buildings can be designed using either method, or both. The former almost inevitably results in far higher levels of fossil fuel use and in turn emissions from buildings to maintain comfort indoors mechanically. The latter method lengthens the annual time in days, weeks and months that the building can operate using local wind/air/temperature energy for heating and cooling at no cost at all to occupants or the environment. The former method shifts the investment emphasis away from the cost of the fabric and design of the building to driving up spending on mechanical systems. The consideration of building performance by architects at an early design stage appears as almost obsolete in too many modern buildings. This is evidenced by international design awards being systematically given to some of the most environmentally damaging buildings of all time that have thin, tight, over-glazed skins framing structures, buildings that are deemed to derive their architectural quality from little more than their sculptural shapes. One eco-architect has described such 'architecture' as little more than 'building hairdressing'.

Does this matter? It did not appear to do so until rather recently. Five years ago we devoted a whole issue of *ASR* (Roaf 2012) to 'Innovative approaches to the natural ventilation of buildings' where some of the imperatives for change in the field were detailed. A range of papers were presented showing how building regulations (Brazil and the Netherlands) are changing to favour natural ventilation, how low tech (ceiling fans) and high tech (mixed mode large and high-rise buildings) work, and how and when people use windows. The papers were widely read and cited but still the building's market did not appear to move much in the natural ventilation direction.

In Roaf, Nicol, and de Dear (2013) (*ASR* 56:1) we returned to the 'Wickedly' complex challenge we face in providing reliable, affordable, comfort to our current and coming generations. Using Japan as an example to highlight the rapid change in

attitude towards comfort that can occur because of 'Events', Tanabe described the comfort implications of the failure of the Fukushima nuclear plant after which all Japanese businesses were made by law to reduce their energy consumption by 20% overnight. We have also shown how incredibly sophisticated is the design of some modern naturally ventilated buildings and also how extraordinarily effective are some of the historic ventilation systems of complex buildings such as the traditional wind towers of the Middle East (McCabe and Roaf 2013).

In Roaf, Nicol, and Rijal (2015), we looked specifically at the escalating problem of providing comfort at high temperatures in an energy insecure future, a changing climate and the unstable economies that appear to reflect the new normal, globally. It concluded that modern designers will have to 'get real' about the *emerging trans-national health threat* posed by buildings in a warming world and change the way they design modern buildings. Not only must their designs reduce energy consumption in, and greenhouse gas emissions from, buildings in practice but designers simply must understand the imperative to reduce overheating in buildings, not least during the power outages that are increasingly common around the world. The message is beginning to get through. As with the introduction of party walls between buildings in London after the Great Fire of 1666, events such as Hurricane Sandy in 2012 are also providing new design imperatives.

From 1997 to 2010, 152 New Yorkers died from heat stroke, often associated with power outages. During heat waves, mortality from other diseases also increases, as the extreme heat exacerbates existing conditions. The New York (quoted from New York Resiliency Task force 2014) Department of Health and Mental Hygiene estimates that deaths from existing conditions, such as cardiovascular and pulmonary disease, increased by 6.5% during 12 prolonged heat waves over that period, representing approximately 1090 additional deaths (New York Building Resiliency Task Force 2014). Not only did buildings get very hot after Hurricane Sandy but during related power outages that also cut off water supplies, many people in high-rise residential buildings continued to use the toilets, but could not, and did not, carry water up to their floors to flush them and in the worst instances recorded air in some buildings became 'putrid'. This resulted in a rapid change in the ordinances relating to water pumping in buildings. Researchers now predict that Manhattan heat deaths will rise by 20% in the 2020s and by 90% by the 2080s as the Climate Warms (Klopott 2013). As the NY resilience Task force notes, the solution in an economically challenging future will not be to turn up the air-conditioning and so increasingly the first line of defence against heat in many climates will be to open the window, safely.

Internationally, evidence proliferates of the importance of providing not only affordable comfort to populations but also shelter against extreme temperatures be they hot or cold. The rising death tolls of the Indian heat waves where increasing numbers die on the streets peaked in May 2015 as temperatures rose above 45°C across the continent. In India when temperatures rise above 40°C, a 'heat wave' refers to a departure of between 4°C and 5°C from the normal temperature while a 'severe heat wave' refers to a departure of more than 6°C.

In Australia even the affluent middle classes are beginning to suffer from regular heat stress. The Australian Bureau of Meteorology confirmed that January 2017 was the hottest month ever recorded, as was February following it. In that city the average maximum temperature for January of 29.6°C beat the previous highest of 29.5°C recorded in 1896. The temperature topped 30°C on 11 days and on 5 days it rose above 35°C, smashing not only all previous records for the month but also for any month since records began in 1858. The real scare came when the New South Wales government issued a warning in February that the power supply system may simply not be able to cope with the peaks of demand and the lights and air-conditioning systems may indeed fail at the hottest times of the day. The implications of these soaring temperatures, even in climatically well-designed buildings are bad enough but in over-glazed, unshaded, poorly oriented buildings, particularly those whose windows do not open they may well prove lethal when the power fails.

Things have to, and are, now beginning to change. Major changes will occur in three ways:

- Buildings will be designed/modified to withstand higher temperatures
- Buildings will increasingly be heated and cooled using natural energy where possible
- People will take back control of the adaptive opportunities in the environments they occupy using a range of manual or mechanical means.

This issue of *ASR* (60.2) has been carefully designed to tell the story of the ways in which these changes are already occurring and the rationale behind them. The elevator pitch for this issue might be summarized as:

If you take an ordinary house, for instance, in the Netherlands (Alders, 150–166) that is well designed for shading, natural ventilation, energy storage and with multiply manually operable adaptive opportunities then modelling shows that not only can nearly two-thirds of heating and cooling energy be saved but also comfort can be maintained even at higher temperatures when temperature peaks can be smoothed down through including passive energy storage in the building materials.

Sassi (167–179) demonstrates that such high energy savings can be achieved in practice when she extensively monitored eight highly insulated dwellings with decentralized, natural ventilation systems for indoor air quality, temperature and relative humidity, garnering comfort responses to validate her findings. She demonstrates that new buildings with natural ventilation do provide adequate comfort and also good indoor air quality for occupants. She looks at the success of decentralized ventilation systems in providing less disruptive and less expensive solutions for high-performance retrofits of the existing housing stock, which is currently overwhelmingly naturally ventilated.

While the two papers above deal with the here and now of housing we have already seen a range of papers showing that even in high latitudes modern buildings in particular are increasingly in danger of overheating in warmer weather (eg. Maivel, Kurnitski, and Kalamees 2014). Brotas (180–191) looks at the criteria from the UK CIBSE TM 52 method for assessing overheating in European buildings and discusses their current applicability to a single dwelling archetype: an urban mid-floor flat. She then models this fairly universal dwelling type and locates it in a range

of European cities to understand causes, extent, characteristics and energy costs of, and remedies for, overheating in the different climates now and in the morphed 2020, 2050 and 2080 climates. The results highlight some problems in practice using simulations tools to evaluate overheating and the fundamental assumptions on which they and TM52 as a predictor of overheating are based.

Simson et al. (192–204) look also at the efficacy of simulations in assessing overheating impacts. They compared three thermal zoning methods for summer thermal comfort assessment: two multi-zone approaches, modelling the whole building or apartment, and a single zone approach, modelling only one room. Simulation results have been evaluated using Coefficient of Variance of the Root Mean Square Errors, Mean Bias Error and average percentage error. They identify some of the methods available, showing that when tested with real data the single zone model provided the best agreement, but typically overestimated overheating because it does not account for the thermal dynamics of the building, through heat dissipation between the zones, and also has limitations in accounting, for example, for cross-ventilation. They conclude that the apartment model based on the multi-zone method gave more realistic results, with little differences to the whole building approach, and can be suggested as an alternative method for more accurate simulations. However, they identify a range of problems in dealing with the comfort effects of air movement and in providing tenants of such flats with ways to predict the probability of overheating discomfort. They do however propose a method with which it is relatively easy to pre-assess an apartment for thermal comfort using only temperature measurements, without having to conduct simulations to prove the existence of overheating problems.

So far we have seen that a well-designed home can, with a low cost and low impact, provide comfort even at high temperatures, using natural ventilation, energy storage and adaptive opportunities, without compromising indoor air quality. Also highlighted has been that many modern buildings are overheating and will increasingly do so in a warming climate. The limitations of current overheating assessment, rating and simulation methods lie in dealing with the performance of passive adaptive features and in managing natural energy flows through the air and fabric of buildings and the potential usefulness of depending on a 'feel for temperatures' and common sense in thinking about how to reduce overheating in buildings at the design stage.

The problem with building models is that they tend to lump the many design factors into one model 'soup' into which go all the physical and mechanical parameters of the building while omitting any realistic manifestation of how people can and might interact with it in reality. The next four papers separate the design process into logical steps one might well follow to make the right decisions about the thermal performance of a building in the right order.

Doctor-Pingel et al. (205–214) provide a wonderful case study of how important is the fundamental form of the building itself to the comfort of people within its lifetime. The case study of Golconde, the first modern reinforced concrete building in India at Pondicherry, South India, still remains as one of the most outstanding examples of climate-responsive buildings in that country. The paper demonstrates how much can be achieved simply through building footprint, planning, orientation, landscaping and the size and placing of openings to ensure comfort in a hot tropical climate using sophisticated natural ventilation paths and controls.

Van den Engel and Kurvers (215–224) demonstrate in model Dutch buildings that even with advanced natural ventilation systems in hospitals and schools, very early design stage thinking is essential to carefully weave into the visible form of the building an invisible natural ventilation network of air pathways through the structure. This is the step that modern designers too often appear to have no understanding of: the importance of the internal energy flows through buildings, be they heat, air or light. The movement of natural energy through the building must surely offer huge opportunities to excel in advanced performance design. The authors also deal with the often ignored issue of turbulence which is also an important comfort parameter, having a positive as well as a negative influence on comfort, and on that needs further work.

Thus designers, having established their desired forms and natural energy flows through their buildings, can also explore a wide range of ways in which building both at the design stage, and in the future, can be adapted to offer comfort in a changing climate. Shiel et al. (225–238) address this issue very nicely with a case study of a suburban house in Adelaide. The ways the house can be improved to reduce its overheating in current and future temperatures are explored and modelled and provide a basis for evaluating the efficacy of different energy and comfort rating and modelling tools in use in Australia. In doing so the value of making both physical and behavioural changes is established and also the real failures in the rating systems to be able to make allowance for the improvements effected.

Here, we get onto the issue of occupant behaviours, perceptions and attitudes skirted around by the previous authors but dealt with head on by Raw et al. (239–248). They undertook a quantitative social survey of 2313 British households to enhance our understanding of how users' needs connect energy use with indoor environmental quality in relation to both energy savings and good indoor environments. They show how comfort must be seen not simply as the thermal state of individuals but in terms of the actions of groups of individuals, in the context of conflicting needs and attitudes. They explore the needs, perceptions and personal interactions of building users under headings of *Other people*, *Comfort* (not only thermal comfort), *Hygiene*, *Resource* and *Ease* with the aim of informing designers better on the wider range of issues and concerns that will influence the building-related experiences of occupants.

Finally, Aynsley and Shiels (249–254) conclude with a think piece that raises a question that bubbles under most of the papers in this special issue: where is the common sense in it all? Why have building design professionals and the construction industry forgotten so much or not learnt from the widespread impact of poor ventilation? Is it because of the single-minded quest for energy efficiency alone rather than overall greenhouse gas emissions reductions? Has the driver for better engineered solutions been at the expense of better designed ones? They reinforce the design imperative underlying the indisputable benefits of adopting more passive and sustainable approaches to ventilation in building design and construction in our warming world. They also raise the challenges posed by the limitation of software and promote new areas to be explored and developed in designing to exploit free comfort gains that can be harvested from elevated airflow for cooling building occupants in summer. Where van den Engels highlights the need for research into the benefit and dis-benefits of turbulence in natural ventilation systems here the authors point to the enhanced cooling that can be garnered from gusts in airflows and the need for research into optimal gusting frequencies from airflow devices for comfort.

This is an exciting *ASR* issue that above all shows how important natural ventilation is, and increasingly must be for the low cost, low impact provision of comfort in buildings. It suggests that responsibility for the design of resilient buildings that can keep people thermally safe in a warming world must be taken by the first-stage designers who shape the immutable form of the building and the air pathways through them. Only then should the second-level decisions be made to provide the 'adaptive' opportunities for heating and cooling the building that may be changed over a day, year or decade. They may include a range of choices from manually to mechanically operated elements such shades, shutters, glazing type and ceiling fans to heating or air-conditioning systems. These secondary systems may all be changed over time but not the fundamental building form itself. Finally, the development and refinement of a design over time must relate to a deeper understanding of how the occupants of the building might use and evolve it through changes in climates, economies, societies and fashions.

It has become clear over the development of this *ASR* issue that the comfort rating tools, regulations and systems at play in the Business-As-Usual, rather backward looking twentieth-century markets of today can often prove an impediment to the development of truly innovative and ultimately more resilient building solutions. For instance new thinking on natural ventilation proliferates but investment priorities and procurement thinking must change too in response, and be modified to reward more common sense approaches that address the need for better basic climatic design for all new buildings. Where does 'modern architecture' fit into all of this? It should be right at the heart of it, as evidenced by the research papers in this issue.

Twenty-five years on from the opening of the Oxford Thermal Comfort Unit the time has come to actually begin to mandate for the universal re-introduction of natural ventilation in buildings, not least so they do remain habitable when the power does fails during the next heat wave. The time has come for Architectural Science and Architectural Sense to make this happen.

Disclosure statement

No potential conflict of interest was reported by the authors.

References

ASHRAE. 2010. *ANSI/ASHRAE Standard 55-2010: Thermal Environmental Conditions for Human Occupancy*. Atlanta: American Society of Heating, Refrigerating and Air-Conditioning Engineers.

BSI. 2005. *BS EN ISO 7730: 2005: Ergonomics of the Thermal Environment. Analytical Determination and Interpretation of Thermal Comfort Using Calculation of the PMV and PPD Indices and Local Thermal Comfort Criteria*. London: British Standards Institution.

BSI. 2007. *BS EN 15251: 2007: Indoor Environmental Input Parameters for Design and Assessment of Energy Performance of Buildings Addressing Indoor air Quality, Thermal Environment, Lighting and Acoustics*. London: British Standards Institution.

Fitch, James Marston. 1966. *American Building: The Historical Forces That Shaped it*. Cambridge, MA: The Riverside Press.

Fitch, James Marston. 1977. *American Building: The Environmental Forces That Shape it*. New York: Schocken Books.

Humphreys, M. A., F. Nicol, and S. C. Roaf. 2015. *Adaptive Thermal Comfort: Foundations and Analysis*. London: Earthscan/Routledge. March 2012 ISBN ISBN: 978-0415691611.

Klopott, F. 2013. *Manhattan Heat Deaths Seen Rising 20% in 2020s as Climate Warm*. BLOOMBERG. Accessed 19 May 2013. http://www.bloomberg.com/news/2013-05-19/manhattan-heat-deaths-seen-rising-20-in-2020s-as-climate-warms.html Accessed 8/2/2017.

Maivel, M., J. Kurnitski, and T. Kalamees. 2014. "Field Survey of Overheating Problems in Estonian Apartment Buildings." *Architectural Science Review* 1–10. doi:10.1080/00038628.2014.970610.

McCabe, C., and S. Roaf. 2013. "The Wind Towers of Bastakiya: Assessing the Role of the Towers in the Whole House Ventilation System Using Dynamic Thermal Modelling." *ASR* 56 (2): 183–194. Accessed 10 October 2012. http://dx.doi.org/10.1080/00038628.2012.723398.

Mendell, M., Q. Lei-Gomez, A. Mirer, O. Seppanen, and G. Brunner. 2008. "Risk Factors in Heating, Ventilating, and air-Conditioning Systems for Occupant Symptoms in US Office Buildings: The US EPA BASE." *Proceedings of Indoor Air* 18 (4): 301–316.

Mendell, M., and G. Mirer. 2009. "Indoor Thermal Factors and Symptoms in Office Workers: Findings from the US EPA Base Study." *Indoor Air* 19 (4): 291–302.

New York Building Resiliency Task Force. 2014. Ensure Operable Windows, Report 26. http://urbangreencouncil.org/sites/default/files/brtf_26-_ensure_operable_windows.pdf.

Nicol, F., M. A. Humphreys, and S. C. Roaf. 2012. *Adaptive Thermal Comfort: Principles and Practice*. London: Earthscan/Routledge. ISBN 978-0-415-69159-8.

Roaf, S. 1993. "Natural Ventilation." In *Proceedings of Profit from Thin Air*, Capenhurst: EA Technology.

Roaf, S. 2012. "Innovative Approaches to the Natural Ventilation of Buildings." *ASR* 55 (1): 1–3. http://dx.doi.org/10.1080/00038628.2012.642190.

Roaf, S., D. Crichton, and F. Nicol. 2009. *Adapting Buildings and Cities for Climate Change*. Oxford: Architectural Press.

Roaf, S., F. Nicol, and R. de Dear. 2013. "The Wicked Problem of Designing for Comfort in a Rapidly Changing World." *ASR* 56 (1): 1–3. http://dx.doi.org/10.1080/00038628.2012.753783.

Roaf, S., F. Nicol, M. Humphreys, P. Tuohy, and A. Boerstra. 2010. "Twentieth Century Standards for Thermal Comfort: Promoting High Energy Buildings." *ASR* 53, 65–77.

Roaf, S., F. Nicol, and H. Rijal. 2015. "Designing for Comfort at High Temperatures." *ASR* 58 (1): 35–38. November 2014, http://dx.doi.org/10.1080/00038628.2014.972069.

Adaptive heating, ventilation and solar shading for dwellings

E. E. Alders ⓘ

ABSTRACT

Calculation of various strategies for the heating of, and the prevention of overheating in, a Dutch standard dwelling that includes (automated) adaptive ventilation systems and solar shading to maintain indoor temperatures at acceptably comfortable temperatures informs this analysis of the costs, impacts and benefits of the use of related control opportunities and mechanisms at play. The energy saving potential of enabling occupants to take advantage of the adaptive opportunities embedded into the dwelling, and discussion of associated cost and benefits of a range of behaviours within the reference dwelling is very high. In the calculations, the total energy saving potential for heating behaviours that take advantage of occupant-driven adaptive behaviours is around 65% of the heating demand for the whole house compared to the saving calculated for the same dwelling controlled by using a standard heating schedule and constant ventilation, which is largely achieved by the use of adaptive controls and fast reaction heating and minimizing ventilation in the heating season. Applying a range of passive cooling strategies, the need for cooling can be eliminated in most situations cancelling the need for the installation of active cooling. It is most effective to use both adaptive ventilation and solar shading.

Introduction

This article is an excerpt of Chapter 6 of the doctoral thesis 'Adaptive Thermal Comfort Opportunities for Dwellings', providing thermal comfort only when and where needed in dwellings in the Netherlands (Alders 2016). In this dissertation, an inventory has been made of adaptive techniques for thermal comfort to create an adaptive thermal comfort system for dwellings, which is defined as follows:

> the whole of **passive and active comfort components** of the dwelling that **dynamically adapts** its settings to **varying user comfort demands** and **weather conditions** (seasonal, diurnal and hourly depending on the aspects adapted), thus providing comfort **only where, when and at the level needed** by the user, to **improve possibilities of harvesting the environmental energy** (e.g. solar gain and outdoor air) when available and storing it when abundant.

Calculating the full potential of adaptive building characteristics

The dissertation calculates the full energy saving potential for heating and cooling for three different profiles of occupancy and three levels of thermal mass (Table 1) of the following adaptive building characteristics:

F_{sol}: solar factor of facade ($q_{rad,gain}/q_{rad,inc}$) – the portion of radiation falling on the entire facade that enters the room as heat

H_{tot} (W/K): total heat transfer coefficient ($H_{ve} + H_{tr}$) – the total heat loss from indoor air to outdoor air per K temperature difference by ventilation and transmission.

Furthermore, the actual thermal comfort demand is determined in time and place to be able to provide heating and cooling only when and where needed according to the definition of the Adaptive Thermal Comfort System. The calculated variants of control are given in Table 2. The values mentioned below are based on the proportions of the reference dwelling of AgentschapNL and in the Dutch climate. These calculations are based on generic physical properties that are regarded to be disconnected (e.g. adapting the glass percentage for the solar factor has no influence on the transmission properties of the facade) to be able to isolate the individual parameters in order to encourage development of new concepts for materials and techniques.

Energy saving potential

Above the energy saving of adaptive heating and minimizing the heat loss factor by high insulation values and ventilation controlled by presence the adaptive approach offers the possibility of increasing the solar gain without causing overheating. This significant amount of energy saving can only be obtained in the theoretical case if the solar factor can be disconnected from the heat loss factor and a 100% of the solar radiation hitting the facade can be used as heat gain ($f_{sol} = 1$). It should be noted that the ultimate energy saving potential for heating and cooling can be higher if there is some form of prediction by an Adaptive Model Predictive Control (Oldewurtel et al. 2012) with anticipation of future comfort demand to prevent the indoor temperature to drop below heating set point at absence due to passive cooling.

Table 1. Calculated cases.

Thermal mass level				Capacitance (J/km^2) (per m^2 floor area)	
Low				80.000	
Middle				165.000	
High				370.000	
Year					
2050 W+				Test reference year by TNO (NEN_5060 2008)	
Room				**Volume (m^3)**	
Living room				85	
Bedroom				42	

Occupancy profiles	Code	People in household	Occupancy rate (%)	Average number of people present	Average activity level
One student	1st	1	10	1.26	2.5
Couple both with job	2w	2	18	1.32	2.1
Couple with 2 small children	4sm	4	46	1.70	2.4

Table 2. Calculated control variants.

Variants	Control $\Phi_{HC,nd}$		Control H_{tot}					Control f_{sol}		
	Timing	Set point	Automation	Frequency	Range U_{opaque} (W/m^{2}K)	Range U_{tr} (W/m^{2}K)	Range Q_{ve} (1/h)	Automation	Frequency	Range f_{sol}
Reference	s	hi/lo	−	−	0.2	1.6	1.25	−	−	$0.6 * f_{glass}$
Adaptive heating	p	ACA	−	−	0.2	1.6	1.25	−	−	$0.6 * f_{glass}$
Max heat loss	p	ACA	−	−	0.2	1.6	30	−	−	0
Min heat loss	p	ACA	−	−	0.1	1.2	min$_{pres}$	−	−	1
Adaptive $H_{tot,hour}$	p	ACA	+	h	0.1−0.2	1.2−1.6	min$_{pres}$-30	−	−	$0.6 * f_{glass}$
Adaptive $H_{tot,day}$	P	ACA	+	d	0.1−0.2	1.2−1.6	min$_{pres}$-30	−	−	$0.6 * f_{glass}$
Adaptive $H_{tot,season}$	P	ACA	+	s	0.1−0.2	1.2−1.6	min$_{pres}$-30	−	−	$0.6 * f_{glass}$
Adaptive $H_{tot,month}$	P	ACA	+	m	0.1−0.2	1.2−1.6	min$_{pres}$-30	−	−	$0.6 * f_{glass}$
Presence $H_{tot,hour}$	P	ACA	−	h	0.1−0.2	1.2−1.6	min$_{pres}$-30	−	−	$0.6 * f_{glass}$
Adaptive $f_{sol,hour}$	P	ACA	−	−	0.2	1.6	min$_{pres}$	+	h	0−1
Adaptive $f_{sol,day}$	P	ACA	−	−	0.2	1.6	min$_{pres}$	+	d	0−1
Adaptive $f_{sol,season}$	P	ACA	−	−	0.2	1.6	min$_{pres}$	+	s	0−1
Adaptive $f_{sol,month}$	P	ACA	−	−	0.2	1.6	min$_{pres}$	+	m	0−1
Presence $f_{sol,hour}$	P	ACA	−	−	0.2	1.6	min$_{pres}$	−	h	0−1
ad $H_{tot,hour}f_{sol,hour}$	P	ACA	+	h	0.1−0.2	1.2−1.6	min$_{pres}$-30	+	h	0−1
ad $H_{tot,day}f_{sol,day}$	P	ACA	+	d	0.1−0.2	1.2−1.6	min$_{pres}$-30	+	d	0−1
ad $H_{tot,season}f_{sol,season}$	P	ACA	+	s	0.1−0.2	1.2−1.6	min$_{pres}$-30	+	s	0−1
ad $H_{tot,month}f_{sol,month}$	P	ACA	+	m	0.1−0.2	1.2−1.6	min$_{pres}$-30	+	m	0−1
pres $H_{tot,hour}f_{sol,hour}$	P	ACA	−	h	0.1−0.2	1.2−1.6	min$_{pres}$-30	−	h	0−1

s, Standard heating schedule; p, presence; hi/lo, set point (21°C) and setback (15°C); ACA, Adaptive Comfort Algorithm (Peeters et al. 2009); −, none; +, automation; H, hourly switch; d, daily switch; s, seasonally switch; m, monthly switch; f_{glass}, percentage of glass in facade.

Cooling

Applying the variable building characteristics as described in the thesis, the cooling can be eliminated in most situations and will be 10% of the initial value at most. The remaining cooling demand will be low enough to cancel the need for installation of active cooling. It is most effective to use both an adaptive heat loss factor and solar factor; however, of the two the solar factor is most effective.

The energy saving potentials of the separate solutions are given in Table 3. The total energy saving potential of the added measures is therefore mentioned in the bottom of Table 3. To increase the thermal storage a PCM or a dynamic thermal storage (HATS) (Hoes 2014) could be applied.

Required values for H_{tot} and f_{sol}

The energy saving potential calculated in the thesis is based on theoretical values for the ranges of the H_{tot} and f_{sol} in a more or less realistic but theoretical range; consequently, the actual energy saving potential can vary according to the actual ranges of the variable building characteristics. Varying the heat loss factor is more feasible with increasing the ventilation because this can more easily reach very high heat loss factors. Developing new techniques for increasing the heat loss factor by transmission is only competitive to ventilation if high conductive properties can be incorporated in new materials or techniques.

In this study, it is shown that the advantage of increasing the solar gain in the heating season is significant. However, varying the solar factor between the values assumed in the calculations from 0 to 1 is not possible with current techniques. Blocking all solar radiation is difficult without blocking all vision out and glazing will always block a percentage of the heat, making it impossible to reach the value of 1 by currently available materials. New techniques could include new materials for advanced radiation transfer or harvesting more solar radiation by increasing the surface of incidence such as applying solar collectors.

Effect of the thermal mass

The most important conclusion about thermal mass in these calculations is that the control of the heat loss factor and the solar factor is significantly more stable with the higher thermal mass. The application of passive cooling by increasing the heat loss factor and/or decreasing the solar gain in case of low thermal mass

Table 3. Summary of energy saving potential of the Adaptive Thermal Comfort System based on the generic calculations.

		4sm		2w		1st		Average	
		Living room	Bedroom	Living room	Bedroom	Living room	Bedroom	Living room	Bedroom
Adaptive heating	Energy saving potential[a]								
	Heating	12%–23%	17%–20%	37%–48%	16%–19%	41%–47%	13%–14%	34%	16%
	Overheating	–	–	–	–	–	–	–	–
Minimized heat loss	ACPH (1/h) $= 0.5 + 0.2*p$			Rc_{op} [km^2/W] $= 10$			U_w [W/km^2] $= 1.2$		
	Energy saving potential[a]								
	Heating	67%–74%	87%–93%	64%–65%	88%–94%	63%–65%	86%–92%	66%	90%
	Overheating[b]	–	–	–	–	–	–	–	–
Adaptive heat loss	ACPH (1/h) $= 0.5 + 0.2*p$			Rc_{op} [km^2/W] $= 2.5$–10			U_w [W/km^2] $= 1.2$–2.5		
	Energy saving potential[a,c] (*)								
	Heating[d]	−5% to 0%	−2% to 0%	−1% to 0%	−3% to 0%!!	−1% 0%!!	−2% to 0%!!	−1%	−1%
	Automated[e]	3%–32%	1%–12%	1%–7%	0%–15%	1%–8%	0%–6%	9%	6%
	Overheating	74%–95%	78%–98%	73%–95%	87%–98%	70%–94%	83%–98%	84%	92%
	Automated[e]	5%–9%	1%–11%	22%–25%	0%–11%	35%–38%	1%–14%	22%	6%
Adaptive solar factor		F_c					0–1		
	Energy saving potential[a,f] (*)								
	Heating[d]	33%–83%	69%–100%	53%–81%	72%–100%	48%–81%	66%–93%	63%	83%
	Automated[e]	36%–52%	96%–100%	49%–68%	72%–99%	46%–74%	66%–90%	54%	83%
	Overheating	97%–100%	92%–96%	99%–100%	91%–95%	99%–100%	96%–98%	99%	95%
	Automated[e]	15%–21%	96%–92%	67%–74%	90%–94%	76%–82%	93%–95%	56%	93%
ACTS	Energy saving potential[a,g]								
	Heating	27%–82%	99%–100%	50%–80%	99%–100%	44%–81%	99%–100%	61%	99%
	Automated[e]	36%–50%	101%–114%	50%–68%	100%–117%	49%–74%	97%–105%	55%	105%
	Overheating	98%–100%	100%	99%–100%	100%	100%	100%	100%	100%
	Automated[e]	2%–7%	3%–24%	20%–43%	2%–23%	33%–59%	2%–28%		

[a] These values are based on the generic calculations of the thesis (Alders 2016) and are based on the reference dwelling of AgentschapNL. The variation in energy saving potential per situation depends on the thermal mass level.

[b] Overheating escalates without additional measures in summer.

[c] For the energy saving potential on cooling by adaptive heat loss coefficient this amount depends on the thermal mass level and (less) on the level of maximum heat loss coefficient.

[d] Heating can be increased by lack of prediction and thermal storage.

[e] Automated control; shows the added energy saving of automated control above non-automated control.

[f] The energy saving potential of the f_{sol} for heating is due to the maximization of the solar gain in the heating season.

[g] The total energy saving potential of all discussed measures compared to the reference situation with average insulation, average solar factor and constant ventilation [1.25 1//h].

–, Not applicable.

leads to an increase in heating demand due to the temperature drop during absence.

In the reference cases the cooling load is significantly lower for higher thermal mass and the heating load does not significantly vary with thermal mass due to the good basic insulation of the dwelling and sufficient solar gain. The energy saving for cooling by the adaptive strategies is higher with high thermal mass more drastically reducing the already much lower cooling demand for higher thermal mass. This can be explained by the fact that the thermal mass causes the heat to be lost to spread more evenly amongst the hours, which means the peaks for heat to be discarded will be much lower.

In case of the North orientation, it can be an option to build with lower thermal mass especially if there is no increase of solar gain available because the solar radiation has significantly less effect on the heating demand, showing a slight increase in heating demand for higher thermal mass.

Effect of occupancy

The control of the thermal comfort system according to the occupancy and the resulting comfort demand is an essential part of the Adaptive Thermal Comfort System as the definition suggests. The difference between the energy saving is apparent in the case of adaptive heating, where high occupancy means less heating demand if a standard heating schedule is used because of higher internal gain the heating demand with control for presence shows less deviation between the occupancy profiles. This means that the energy saving is larger with lower occupancy level. Furthermore, the control algorithms are significantly less accurate in case of the occupancy profile 4_sm with high occupancy. This effect decreases with higher thermal mass because of the slower temperature decrease.

Additionally, the occupancy has a great effect on the effectiveness of non-automated systems in case of overheating prevention. This emphasizes the need for automated overheating protection by especially solar shading in case of larger periods of absence during the day.

Heat up and cool-down speed

Heat up and cooling down speed are important characteristics of heating systems. In these calculations, it is assumed that the heating power is unlimited and there is no limit to the speed. However, the calculated required heat up speed is largely in a practical range especially if the temperature is stabilized around the comfort temperature by thermal mass, insulation and the adaptive heat loss factor and solar factor. Nevertheless, heat up and cool-down speed are serious considerations for the concept of adaptive heating.

Table 4. Summary of occupancy characteristics of occupancy profiles to be used in the calculations

Characteristic	Composition	LIVING ROOM			KITCHEN			MASTER BEDROOM			BEDROOM 2 /OFFICE			BEDROOM 3			Not at home [%]
		Occupancy rate [%]	Average occupancy	Average activity	Occupancy rate [%]	Average occupancy	Average activity	Occupancy rate [%]	Average occupancy	Average activity	Occupancy rate [%]	Average occupancy	Average activity	Occupancy rate [%]	Average occupancy	Average activity	
profile																	
1_st	1 person household, student	10	1.26	2.5	2	1	3.25	33	1	1	8	1	2	–	–	–	53
1_soc	1 person household, social with much visit	17	1.19	2.0	11	1	3.30	34	1	1	17	1	2	–	–	–	6
2_h	2 person household, at least one with work from home or no job	54	1.35	2.0	14	1	3.45	62	1.39	1	54	1	2	–	–	–	0
2_w	2 person household, both with job	18	1.32	2.1	8	1	3.25	39	1.73	1	1	1	2	–	–	–	41
4_sc	4 person household, school going children	30	1.69	2.0	8	1.55	3.70	40	1.81	1	30	1	1	46	1	1	28
4_sm	4 person household, two children under the age of 5	46	1.70	2.4	15	1.35	3.20	37	1.77	1	43	1	1	34	1	1	16

*household types based on the household types reported in the Time Use Survey (NIWI 2002)
NIWI. 2002. "Tijdsbestedingsonderzoek 2000 TBO'2000." In.: Netherlands Institute for Scientific Information Services.

Practical solutions for standard housing in the Netherlands

In the previous section for both the solar factor (f_{sol}) and the specific heat loss coefficient (H_{tot}) optimal ranges and switching frequencies have been found and the full energy saving potential is very large. However, the approach was to disconnect their characteristics from (existing) techniques not to be hindered by technical limitations beforehand. In this paper, the techniques of the inventory for adaptive techniques made in the dissertation are linked to the conclusions of the generic physical calculations to be translated to practical concepts and constraints for these concepts. Some of the concepts are evaluated for applicability and efficiency in a Reference Dwelling of AgentschapNL in various occupancy scenarios and with high and low thermal mass. It reconnects the previous conclusions to design practice and gives guidelines how to design an Adaptive Thermal Comfort System.

Researched concepts for a standard Dutch dwelling (reference dwelling of AgentschapNL)

Context: the occupant
The dwelling of the example concepts will be a family home which can be occupied by an undetermined family. The calculations to assess the energy saving potential of the Adaptive Thermal Comfort System will be made with the same three occupant profiles used in the generic calculations with the occupancy profiles and characteristics as shown in Table 4 derived from a Time Use Survey conducted by CBS (NIWI 2002). In this example concept, the spatial layout of the attached reference dwelling by AgentschapNL (DGMR 2006) is calculated, which is regarded as a standard Dutch dwelling. Figure 1 shows the floor plans of this dwelling. The living room is oriented South and has large windows to be able to optimally allow in or block solar radiation. The bedrooms are oriented South or North and have smaller windows. Figure 2 shows the facades of the dwelling.

Context: the weather
The dwelling for the example concepts will be situated in the Netherlands, which has a dominant heating demand (in winter). However, as concluded in the thesis the summers are hot and sunny enough to cause overheating problems if the dwelling is designed to predominantly save heating energy by insulating the dwelling very well without extra measures in summer. Figure 3 shows a summary of the incoming solar radiation per month for the reference dwelling of AgentschapNL (DGMR 2006) oriented with the back facade to the South and the maximum, average and minimum ambient temperature per month for the test reference year of 2050 W+ (NEN_5060 2008; KNMI 2014).

Adaptive solar shading
In the example, with a G-value of the glass of 0.6 (HR++), the maximum range of solar factor for the South facade of the living room is 45–0%. For the rest of the facades this is 25–0%. In the TRNSYS simulation, the f_{sol} of the North facade will be varied; however, it is expected not to have much effect because

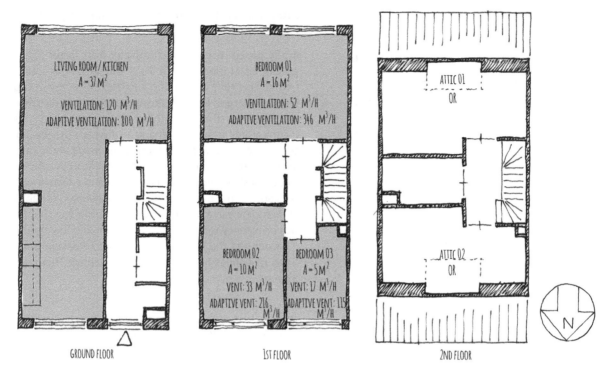

Figure 1. Floor plans of dwelling used for case study.

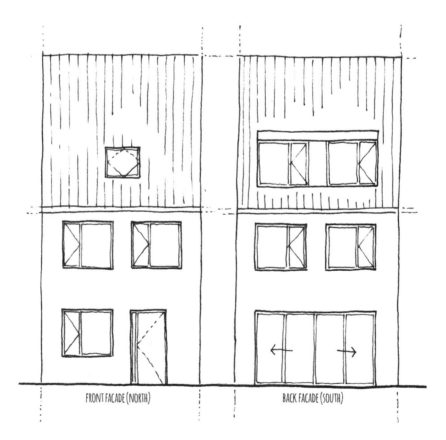

Figure 2. Facades of dwelling used for case study.

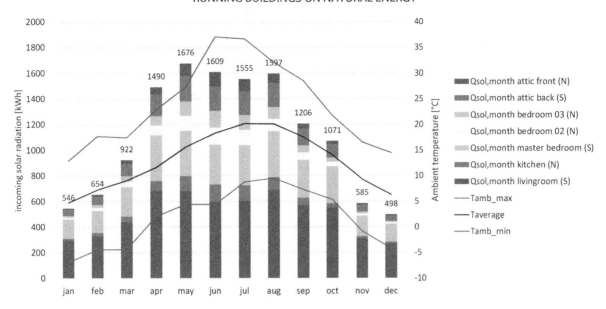

Figure 3. Incoming solar radiation through the windows for the whole house together with the average ambient temperature per month (2050 W+).

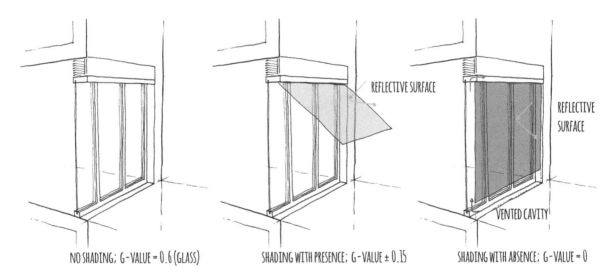

NO SHADING; G-VALUE = 0.6 (GLASS)   SHADING WITH PRESENCE; G-VALUE ± 0.15   SHADING WITH ABSENCE; G-VALUE = 0

Figure 4. Concept for varying the f_{sol} combined screens and awnings.

of the little solar radiation during the summer season. A combination between screens and controllable awnings is chosen in the form of awnings with a reflective surface on the outside that can also be placed parallel to the glass with a cavity that can be vented to prevent overheating of the screen (Figure 4). With this solution (otherwise it will be completely dark when people are home) it is possible to reach the maximum range for solar shading; however, it should be noted that with this setting, also the visual light and view out are blocked. In practice this might not be desirable. However, the beam radiation can be blocked and shading values of 0.15 can easily be reached with the awning setting. Therefore, during presence the maximum solar shading will be a G-value of 0.15. Furthermore, when the screen is parallel to the glass, a cavity between the glass and the screen should be created to ventilate the heat from the surface of the screen that will be heated up by the sun, to make sure no significant solar radiation will 'leak' into the space.

Adaptive ventilation

For the ventilation for temperature control, considerable openings should be present, as shown in the example in Figure 5, which will be simulated in the TRNSYS variants for assessment. The control strategies for natural ventilation assessed in the example concepts are:

- Pressure controlled inlet (constant inlet flow, exhaust controlled mechanically)
- CO_2 controlled ventilation (minimal ventilation, exhaust controlled mechanically)
- Temperature and CO_2 controlled ventilation (minimal ventilation, exhaust controlled mechanically)

Thermal mass

In general, it is preferred to build in high thermal mass as also becomes clear from the previous calculations. In the example, two levels of thermal mass will be researched: low and high to

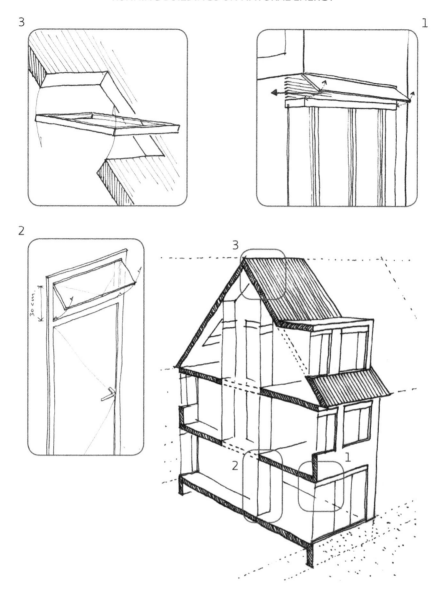

1: Vents above windows of 30 cm height controlled for adaptive ventilation for passive cooling

2: Vents above internal doors of 30cm height opened together with vents above windows in the corresponding roomto enhance airflow through the building

3: Rooflight in staircase opened with either window vent in the house to enhance airflow through the building

Figure 5. Concept for varying the natural ventilation for passive cooling by purge ventilation in the TRNSYS simulations (combined with base minimized ventilation natural supply, mechanical exhaust).

reassess the influence of the thermal mass on the performance of the Adaptive Thermal Comfort System (Table 5).

Simulated variants

The design example is calculated in TRNSYS with the different cases (Table 5) and the control variants shown in Table 6. The overheating is measured in degree hours above cooling set point. At presence, this is the upper limit of the comfort bandwidth and at absence, this is 30°C. The design will be regarded as a whole house and all the performances of the rooms will be summed or averaged.

Energy use of the reference situations

First, the reference situations for all combinations of all cases are calculated, with assumptions shown below in Table 7.

Figure 6 shows the annual heating demand for each combination of occupancy profile and thermal mass variant together with the distribution of that energy demand over the months of the year. Figure 7 shows the overheating in degree hours. The overheating numbers in the graphs are the average of overheating degree hours of all used rooms in the occupancy profiles. In the two-person household the second bedroom is used sparsely and the third bedroom is unused. In the one-person household both the second and third bedrooms are unused.

Table 5. Calculated cases in TRNSYS.

CASES					
Thermal mass			Capacitance (kg/m^2) (per floor area)		
Low			80.000		
High			370.000		
Year					
2050 W+			Test reference year, warm		
Occupancy profiles	Code	People in household	Occupancy rate	Average amount of people present	Average activity level
1 student	1st	1	10%	1.26	2.5
Couple both with job	2w	2	18%	1.32	2.1
Couple with 2 small children	4sm	4	46%	1.70	2.4

Table 6. Calculated control variants in TRNSYS.

Control variants (all variants except for the presence controlled variants (e&f) are calculated with natural ventilation supply as well as with mechanical ventilation supply with heat recovery)		Heating and cooling		Ventilation				Shading	
Variant	Code	s/p	hi-lo/ACA	+/−	a/p	Base ventilation rate (1/h)	0/1	a/p	0/1
Reference	1_ref	s	hi/lo	+	−	1.25	0	−	0
Adaptive heating	1a_ref	p	ACA	+	−	1.25	0	−	0
Minimized ventilation	1b_ref	p	ACA	+	−	min$_{pres}$	0	−	0
Minimized ventilation nc	1c_ref	p	ACA	−	−	min$_{pres}$	0	−	0
Adaptive ventilation	2b_ventdyn	p	ACA	+	a	min$_{pres}$	0–1	−	0
Adaptive ventilation nc	2c_ventdyn	p	ACA	−	a	min$_{pres}$	0–1	−	0
Presence ventilation	2e_ventdyn	p	ACA	+	p	min$_{pres}$	0–1	−	0
Presence ventilation nc	2f_ventdyn	p	ACA	−	p	min$_{pres}$	0–1	−	0
Adaptive ventilation	3b_soldyn	p	ACA	+	−	min$_{pres}$	0	a	0–1
Adaptive ventilation nc	3c_soldyn	p	ACA	−	−	min$_{pres}$	0	a	0–1
Presence ventilation	3e_soldyn	p	ACA	+	−	min$_{pres}$	0	p	0–1
Presence ventilation nc	3f_soldyn	p	ACA	−	−	min$_{pres}$	0	p	0–1
Adaptive ventilation	4b_dyn	p	ACA	+	a	min$_{pres}$	0–1	a	0–1
Adaptive ventilation nc	4c_dyn	p	ACA	−	a	min$_{pres}$	0–1	a	0–1
Presence ventilation	4e_dyn	p	ACA	+	p	min$_{pres}$	0–1	p	0–1
Presence ventilation nc	4f_dyn	p	ACA	−	p	min$_{pres}$	0–1	p	0–1

s/p, s = standard heating schedule/p = presence controlled (adaptive heating); hi-lo/ACA, hi-lo = set point and setback/ACA Adaptive Comfort Algorithm (Peeters et al. 2009); +/−, + = cooling/−− = no cooling, calculating overheating in degree hours; a/p, a = adaptive control (automated)/p = operated with presence (non-automated); 0/1 (ventilation), 0 = no extra ventilation/0–1 = extra ventilation on or off; 0/1 (shading), 0 = no solar shading/0–1 = shading on or off.

Table 7. Initial values for the reference situations.

INITIAL VALUES		
Operation		
Ventilation (1/h)	1.25	(0.9 l/sm^2)
Schedule living room	**Heating set point (°C)**	**Cooling set point (°C)**
0:00–6:00	15	30
6:00–23:00	20.5	25
13:00–0:00	15	30
Boundary	**Heating set point (°C)**	**Cooling set point (°C)**
0:00–6:00	18	25
6:00–23:00	15	30
13:00–0:00	18	25
Boundary	**Equipment (W/m^2)**	**People (W)**
Absence (occupancy schedule)	1	–
Presence (occupancy schedule)	10	75 * (amount of people)

The heating demand does not vary much between high thermal mass and low thermal mass in Figure 6 and is slightly higher for the low thermal. Furthermore, the heating is significantly higher the lower the occupancy pattern, which is caused by the equal heating hours for each profile but more internal gain for the profiles with higher occupancy.

The overheating is significantly higher for lower thermal mass and significantly higher for higher occupancy rates, as Figure 7 shows. The rooms that are not used in the profiles show no overheating degree hours because the hours are only measured at presence.

Results and discussion

Adaptive heating

The next step is researching the energy saving potential for adaptive heating; providing heating only when and where needed at the level needed as opposed to fixed day and night temperatures. The comfort temperature will be calculated by the adaptive comfort algorithm method and the heating will only function at the presence of the occupants or if the temperature falls below the setback temperature of 15°C. In practice, the temperature will never fall below this 15°C because the insulation is high enough to prevent this. Figure 8 shows the energy saving potential for adaptive heating compared to the reference situation; absolute yearly energy consumption (kWh) for adaptive heating and relative to the reference situation with scheduled heating (%). Figure 9 shows the overheating in degree hours absolute and relative to the reference situation (Figures 6 and 7). The overheating numbers in the table are the average of overheating degree hours of all used rooms in the occupancy

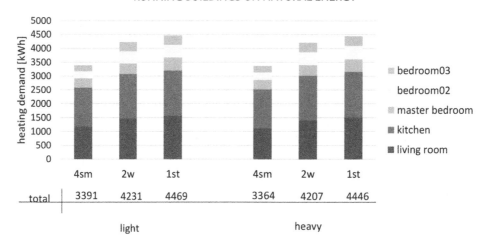

Figure 6. Energy use for heating in kWh in the whole house in the reference situations with fixed ventilation and day and night setting for heating.

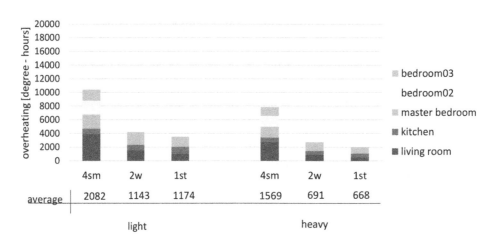

Figure 7. Overheating in degree hours in the whole house in the reference situations with fixed ventilation and day and night setting for heating.

profiles. It should be noted that in the simulations it is assumed that the heating and cooling power is unlimited and has a response time equal to the calculated time step. This means that inertia of the heating delivery system by its thermal mass is not regarded. Systems in practice could therefore be less effective the slower they are.

As becomes clear by Figure 8, adaptive heating can achieve a significantly better energy performance for the whole house. The energy saving is very little (around 5%) for the family with two children (4sm) because their occupancy profile mostly resembles the standard heating times set by the thermostat and the one-person household with the lowest occupancy rate and the least people in the household can gain almost 40% energy saving by the adaptive approach, which is a similar amount. The overheating (Figure 9) is similar to the reference situation showing a slight decrease in the light variant. This is because extra heating can cause a slight cooling demand later on depending on the occupancy of the rooms and incoming (extra) heat by solar radiation. It is remarkable that the rooms that are not in use still show a heating demand. This can be explained by the fact that the heating set point at absence in every room is still 15°C as opposed to 16–180°C at presence in the bedrooms and the ventilation is standardized at an ACPH of 1.25 1/h, which is quite high.

Minimized ventilation

The energy saving for heating by minimizing the ventilation is calculated taking into account the fact that this significantly increases overheating. Figure 10 shows the absolute energy consumption (kWh) for heating by minimized ventilation and relative to the situation of adaptive heating (%) and Figure 11 shows the overheating in degree hours above cooling set point in house reference situations with minimized ventilation. The overheating numbers in the table are the average of overheating degree hours of all used rooms in the occupancy profiles.

From Figure 10, it can be concluded that the energy saving of minimizing the ventilation in the heating season is very effective (almost 50%) and most effective with the most heating hours, which is the case of the profile with the couple with two children (4sm) in addition to the energy saving by adaptive heating. It is now noticeable that the unoccupied bedroom shows a higher decrease in heating demand than the living room, kitchen and the master bedroom because now the ventilation is always low at an ACPH of 0.5 1/h.

However, Figure 11 shows that the overheating problems are significant if no counteractions are taken. This problem aggravates with rising occupancy because the overheating degree

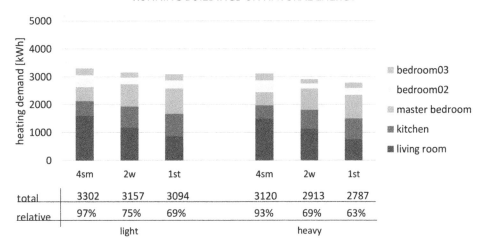

Figure 8. Energy use for heating in the whole house for adaptive heating absolute (kWh) and relative to reference situation (Figure 6) with fixed heating schedule (%).

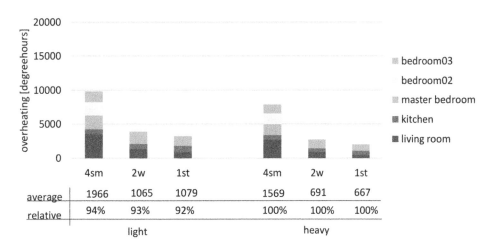

Figure 9. Overheating in degree hours in the whole house for adaptive heating absolute and relative to reference situation (Figure 7) with fixed heating schedule (%).

hours are only calculated for the presence hours. Furthermore, it shows that the overheating is significantly higher for lower thermal mass. The energy saving potential for heating in case of minimized ventilation is slightly lower for higher thermal mass; however, the remaining heating demand is still lower for higher thermal mass.

Adaptive ventilation by operable vents above the windows

To benefit from the energy saving for heating with minimized ventilation without the disadvantage of the overheating problems, the ventilation can be increased whenever there is a surplus of heat in the dwelling. Figure 12 shows the remaining

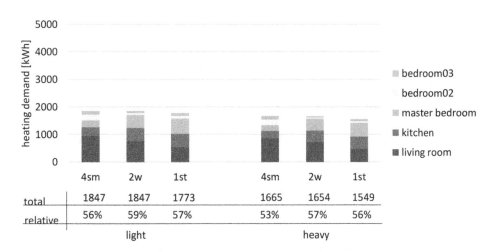

Figure 10. Energy use for heating in the whole house for minimized ventilation absolute (kWh)] and relative to reference situation (Figure 8) with adaptive heating (%).

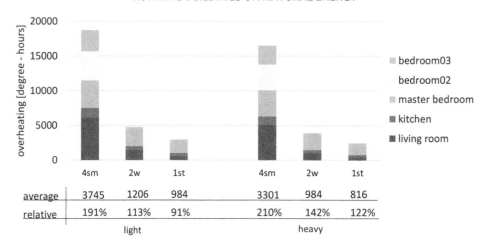

	4sm	2w	1st	4sm	2w	1st
average	3745	1206	984	3301	984	816
relative	191%	113%	91%	210%	142%	122%
		light			heavy	

Figure 11. Overheating in the whole house for minimized ventilation absolute (degree hours) and relative to reference situation (Figure 9) with adaptive heating (%).

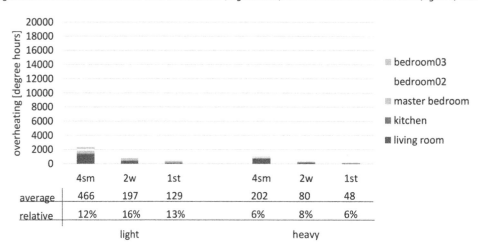

	4sm	2w	1st	4sm	2w	1st
average	466	197	129	202	80	48
relative	12%	16%	13%	6%	8%	6%
		light			heavy	

Figure 12. Remaining overheating with adaptive ventilation absolute (degree hours) and relative to minimized ventilation (%) (Figure 11).

overheating in summer by adaptive ventilation absolute (degree hours) and compared to the reference situation of Figure 11 (%). The overheating numbers in the table are the average of overheating degree hours of all used rooms in the occupancy profiles.

As can be seen from Figure 12 compared to Figure 11, the overheating has diminished significantly but it can still be desirable to have an additional cooling system especially for the low thermal mass variant for the family with small children (4sm) with around an average of 466 degree hours left in the living room, kitchen and three bedrooms on the first floor but it will have significantly less energy demand. In a terraced dwelling usually the staircase will be able to provide some stack effect to enhance the extraction of air placing a controllable opening in the top of the staircase. Adding a Venturi-shaped chimney exit will increase the ventilation more. To make it adaptable the openings should be closable (Figure 13).

Adaptive solar gain

The overheating problem caused by the minimum ventilation can be counteracted by blocking solar radiation as well. Figure 14 shows the overheating left in case of adaptive solar gain strategy absolute in degree hours and compared to the reference situation of minimized ventilation (%).

From Figure 14, it can be concluded that the problems that occur in summer due to the minimization of ventilation can be significantly decreased by blocking unwanted solar radiation, more so than with adaptive ventilation. In this situation, not applying cooling will only lead to significant overheating problems in case of a light construction with occupancy by the 4-person household (4sm) with an average of 277 degree hours left in the living room, kitchen and three bedrooms on the first floor. Countering overheating with solar shading is around twice as effective as adaptive ventilation.

Adaptive heating, ventilation and solar gain combined

Applying both adaptive ventilation and solar gain is the most effective way of energy saving for the dwelling because this can result in almost eliminating of the cooling demand with less ventilation rates required so the openings for ventilation can be smaller in theory because most excess heat is already blocked by the shading. In the calculations of this paper, the same vent openings are used but they will be equipped significantly less because most of the excess heat is already blocked by the solar shading. This will result in less frequently used vent openings and lower average ventilation rates.

Figure 15 shows the overheating left in case of adaptive ventilation and solar gain strategy absolute in degree

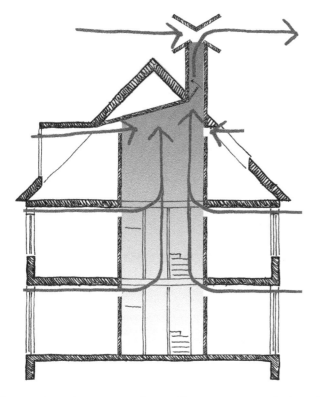

Figure 13. Using the staircase to enhance air flow by extra extraction by the stack effect and the venture-shaped chimney exit.

hours and compared to the reference situation of minimized ventilation (%).

From Figure 15, it can be concluded that as expected the combination of the two measures will almost eliminate virtually all demand for cooling. In case of the light dwelling occupied by the four-person household, 2% of the original overheating is left. With 58 degree hours left on average in the whole house, it is unlikely that active cooling will be required.

Even though the separated measures of dynamic ventilation control and solar shading can prevent most overheating of the home and thus cooling demand, there are clearly benefits in a combination. The vents above the windows could be significantly smaller, which decreases the risk of draught and uses less space in the facade. Furthermore, the smaller the openings the

easier it is to make them burglary proof. Additionally, less effective solar shading could be applied as well, which increases the possibilities for materials and techniques and will allow for saving in the cost of the shading. In practice these measures can be optimized together. Furthermore, the need for shading on the North facade can be omitted totally, which can significantly save costs.

Automation versus manual operation

The energy saving potentials of the three strategies in the past sections all assume there is automated control of the settings that choose the right setting for every situation, even when the occupants are not present. This requires advanced domotics with moving mechanical parts to change the position of the window vents and solar shading which can be vulnerable to break down and intentionally inflicted damage and they can be very costly. To be able to make an informed decision about the level of automation chosen in a design, the energy saving potential of all the measures is calculated if applied only during the presence of the occupants as if they could adjust the ventilation and solar shading by hand preferably with an intelligent system of sensors that gives a warning when something should be adjusted.

Figure 16 shows the remaining overheating with adaptive ventilation only when present for all combinations of occupancy profiles and thermal mass in degree hours and relative to the reference situation of minimized ventilation (%).

Figure 17 shows the remaining overheating with adaptive solar shading only when present for all combinations of occupancy profiles and thermal mass in degree hours and relative to the reference situation of minimized ventilation (%).

Figure 18 shows the remaining overheating with adaptive ventilation and solar shading only when present for all combinations of occupancy profiles and thermal mass in degree hours and relative to the reference situation of minimized ventilation (%).

As evidently becomes clear from Figures 16 to 18, the energy saving potential of automation is significant especially for solar shading. In the bedrooms there is no decrease in overheating with presence controlled solar shading because the sun will only shine significantly in the non-occupied hours when the solar

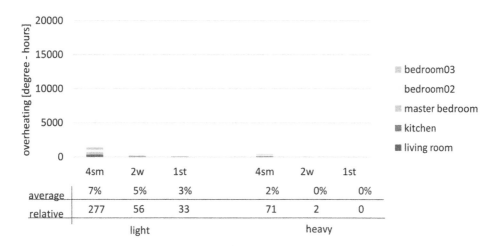

	4sm	2w	1st	4sm	2w	1st
average	7%	5%	3%	2%	0%	0%
relative	277	56	33	71	2	0
	light			heavy		

■ bedroom03
bedroom02
■ master bedroom
■ kitchen
■ living room

Figure 14. Remaining overheating with adaptive solar gain absolute (degree hours) and relative to minimized ventilation (%) (Figure 11).

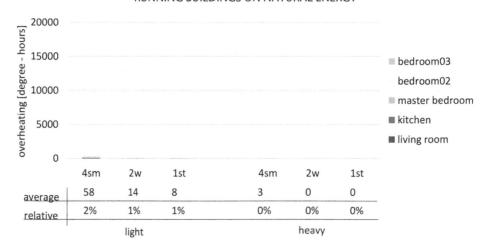

Figure 15. Remaining overheating with adaptive ventilation and solar gain absolute (degree hours) and relative to minimized ventilation (%) (Figure 11).

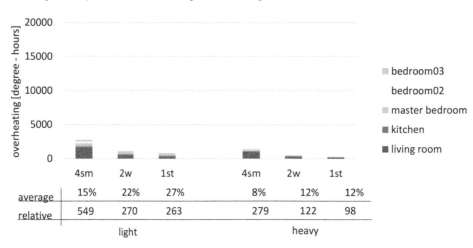

Figure 16. Remaining overheating with adaptive ventilation only when present for all combinations of occupancy profiles and thermal mass absolute (degree hours) and relative to the minimized ventilation variant (%).

shading is not operated. For the adaptive ventilation, the differences are significant but considerably less prominent especially in case of the profile with the highest occupancy rate. The more the people are present, the less the difference between automated and presence operated and the difference is higher with high thermal mass than low thermal mass. Applying solar

shading only when the occupants are present will be significantly less effective than automated solar shading, leaving overheating levels almost similar to no solar shading with a decrease in effectiveness of up to 90%. If both measures can be applied only during presence the overheating is still considerable; at most for the family with young children (4sm) there will be an

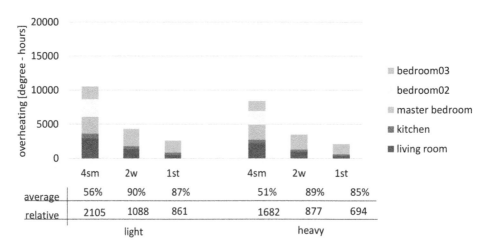

Figure 17. Remaining overheating with adaptive solar shading only when present for all combinations of occupancy profiles and thermal mass absolute (degree hours) and relative to the minimized ventilation variant (%).

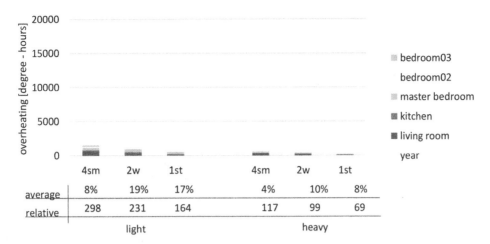

Figure 18. Remaining overheating with adaptive ventilation and solar shading only when present for all combinations of occupancy profiles and thermal mass absolute (degree hours) and relative to the minimized ventilation variant (%).

average of 298 degree hours left in the low thermal mass variant and 117 degree hours in the high thermal mass variant. Nevertheless, this is still a decrease of around 90% compared to the reference situation with minimized ventilation, which is enough to consider omitting the automation, which can be costly and might not be preferred by the users.

Auxiliary energy

Regarding energy saving for (new) techniques, it is important to also incorporate the energy for operation of fans and control systems, the so-called auxiliary energy. In case of the natural ventilation, no extra energy for operation fans is needed, only the fan energy for the mechanical extraction, which is less the less ventilation is applied. Extra auxiliary energy is needed for the operating system and communication as well as the automation of the ventilation openings and solar shading. Communication nowadays is present in most systems, wireless or wired. Most homes will have a network present at which the system can be connected to the Internet to communicate with the components of the systems and also enable the occupant to control the settings at a distance via Internet. It is not expected that the extra electricity for this communication will be anywhere near the energy saving it provides. To make sure no excess energy is spent, the components on the facade can be provided with photovolotaic cells that will provide the little energy needed to operate and they can contribute to the electricity needed to communicate with the system. In this design the solar cells can be applied on the hatch of the window vent. Furthermore, solar cells can be added to the solar screen; however, they would only operate when solar shading is needed. With the further development of the components and the system, these aspects should be taken into account.

Conclusions

Requirements for the Adaptive Thermal Comfort System

This paper described a practical solution for the Adaptive Thermal Comfort System for dwellings as researched in the doctoral thesis *Adaptive thermal comfort opportunities for dwellings; Providing thermal comfort only when and where needed in dwellings*

in the Netherlands (Alders 2016). A comparison of these techniques shows that there is not one perfect system or solution and per project all considerations should be made to design an optimal system. It should be noted that there are numerous techniques to construct an Adaptive Thermal Comfort System and the energy saving potential depends on various aspects and the collaboration between the applied techniques in specific scenarios. some of these techniques are already available and some are in various stages of development. In this paper the possibilities of an Adaptive Thermal Comfort System for the near future is researched. Furthermore, the important aspects to consider about the comfort demand and natural thermal energy supply by the weather and how they should be combined are once more stressed to show the approach needed to design the Adaptive Thermal Comfort System. The conditions for the spatial layout of the dwelling to enable effectiveness of the adaptive measures is described as well as the aspects needed to be considered for control of the systems.

Preconditions for the effectiveness of an Adaptive Thermal Comfort System

Orientation
To optimize effectiveness of the adaptive solar gain both for saving energy for heating as well as cooling by allowing maximum solar radiation in and blocking maximum solar radiation, the rooms with the highest heating demand and/or very variable comfort demand should be oriented in the direction where most solar radiation comes from, considering the time of day of the highest heating demand.

Ventilation
To ensure the effectiveness of the adaptive ventilation the layout should not hinder the air flow through the building. In case of a heating demand this free air flow can be temporarily disabled by (automatically) closing doors and vents. From the calculations it becomes clear that the concept (Figure 5) with large operable vents above the windows of 30 cm together with operable vents above the internal doors and an additional opening in the roof in the stair case has significant reduction of overheating. To be able to reduce the overheating more by ventilation,

this paper suggests a concept to enhance the ventilation more by stack ventilation and a Venturi-shaped chimney (Figure 13). These will have an additional advantage of lowering the need for fan energy for the mechanical exhaust.

Automation

An optimal Adaptive Thermal Comfort System is automated and therefore the system should be provided with sufficient information about the weather and occupant at the right time. For this a design should be made for the sensors and information transfer to the control unit.

This automation also implies communication between the control unit and the end units, which should be able to operate automatically by a signal without interaction with the user. In this design, it is crucial to consider the acceptance of the user of this fully automated system.

Composition of the adaptive components of an Adaptive Thermal Comfort System in a standard reference dwelling in the near future

Adaptive heating: heating only where and when needed at the level needed by the user.

Automated solar shading: solar shading controlled to block solar radiation when needed to prevent overheating and allowing maximum amount of solar radiation in the heating season to decrease the heating demand.

Automated adaptive ventilation: ventilation (preferably by natural ventilation to save fan energy and space for ducts) controlled to discard excessive heat when needed to prevent overheating and minimize ventilation for fresh air in the heating season to decrease the heating demand.

Figure 19 shows a visualization of a full concept of applying an Adaptive Thermal Comfort System into a standard reference dwelling with techniques nowadays available.

Conclusions: energy saving potential of the Adaptive Thermal Comfort System in a standard reference dwelling

In this paper, the conclusions made by the preliminary calculations in the thesis are verified with concepts developed as examples for an Adaptive Thermal Comfort System to be applicable in a current design. It shows that minimizing the ventilation in winter can save almost half of the energy used for heating and that by adaptively blocking the solar radiation and raising the

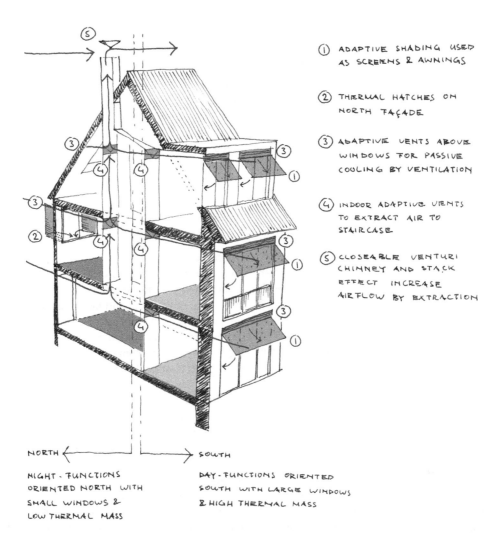

① ADAPTIVE SHADING USED AS SCREENS & AWNINGS

② THERMAL HATCHES ON NORTH FAÇADE

③ ADAPTIVE VENTS ABOVE WINDOWS FOR PASSIVE COOLING BY VENTILATION

④ INDOOR ADAPTIVE VENTS TO EXTRACT AIR TO STAIRCASE

⑤ CLOSEABLE VENTURI CHIMNEY AND STACK EFFECT INCREASE AIRFLOW BY EXTRACTION

NORTH ← | → SOUTH

NIGHT-FUNCTIONS ORIENTED NORTH WITH SMALL WINDOWS & LOW THERMAL MASS

DAY-FUNCTIONS ORIENTED SOUTH WITH LARGE WINDOWS & HIGH THERMAL MASS

Figure 19. Elements of an Adaptive Thermal Comfort System in a standard Dutch dwelling.

Table 8. Summary of energy saving potential of the Adaptive Thermal Comfort System based on the generic calculations.

		4sm		2w		1st		
		Light	Heavy	Light	Heavy	Light	Heavy	Average
Adaptive heating	Energy saving potential[a]							
	Heating	3%	7%	25%	31%	31%	37%	22%
	Overheating	–	–	–	–	–	–	–
Minimized ventilation	ACPH (1/h)	$0.5 + 0.2*p$						
	Energy saving potential[a]							
	Heating	44%	47%	41%	43%	43%	44%	44%
	Overheating[b]	–	–	–	–	–	–	–
Adaptive heat loss coefficient	ACPH (1/h)	0.5–10						
	Energy saving potential[a]							
	Heating	–	–	–	–	–	–	–
	Overheating	88%	94%	84%	92%	87%	94%	90%
	Automation[c]	−2%	−2%	−6%	−4%	−14%	−6%	−6%
Adaptive solar factor	Fc	$0-G_w * f_g$						
	Energy saving potential[a]							
	Heating	–	–	–	–	–	–	–
	Overheating	93%	98%	95%	100%	97%	100%	97%
	Automation[c]	−49%	−49%	−86%	−89%	−84%	−85%	−74%
ATCS	Energy saving potential[a,d]							
	Heating	59%	61%	66%	68%	68%	71%	65%
	Overheating	98%	100%	99%	100%	99%	100%	99%
	Automation[c]	−6%	−3%	−18%	−10%	−16%	−8%	−10%

[a]These values are based on the calculations with the assumptions in the thesis (Alders 2016) and are based on the reference dwelling of AgentschapNL. The variation in energy saving potential per situation depends on the thermal mass level.
[b]Overheating escalates without additional measures in summer.
[c]In this row, the negative percentage represents the decrease in effectiveness against overheating if the adaptive measure is not automated.
[d]The total energy saving potential of all discussed measures compared to the reference situation with average insulation, average solar factor and constant ventilation (1.25 1/h).
–, Not applicable.

ventilation can prevent overheating that can occur for an important part as a result of this minimized ventilation. This shows that with techniques already available, a dwelling can be highly reactive to the changes in the weather, optimizing the thermal heat balance for the majority of occurring situations. It also shows that the measures to vary the ventilation and the solar factor are well within the range of possibilities already available with common techniques. The ventilation openings above the window of approximately 30 cm can create a high enough ventilation rate together with openings above the internal doors of also 30 cm to prevent the dwelling from overheating. The ACPH can rise up to 12 1/h; however, this is a peak value that only occurs less than 5% during the summer months. The prevention of overheating can be enhanced by additional measures to propagate natural air flow for ventilation as well as minimizing the needed for fan energy for mechanical extraction.

The need for cooling can be effectively diminished using the proposed flexible measures. Practically, this means that there should be no need for installing active cooling in the dwelling and thus an energy saving potential of 100% can be reached. For saving energy on heating the situation of minimized ventilation with heat recovery is clearly the best option. The total energy saving compared to fixed natural ventilation is dramatic (almost 50% energy saving). However, as seen before applying all measures for energy conservation in the heating season needs counteractions in the cooling season to prevent overheating. Table 8 shows a summary of the characteristics and energy saving potential for all separate measures and in the end of all measures together. In all cases the remaining heating demand is less than half of the original demand ranging from 41% to 29%. The loss in effectiveness of the adaptive solar gain is

markedly decreased without automation with an average of 75% less energy saving, whose effect is most apparent with lower occupancy rate. The energy saving potential for the combined measures without automation drops with 30% on average, making the automation crucial for the Adaptive Thermal Comfort System.

Smart application of operable windows and solar shading will eliminate the need active cooling in the Dutch residential sector while applying high energy saving for heating demand as the Adaptive Thermal Comfort System shows.

Acknowledgements

The Dutch institute of AgentschapNL partly funded this PhD research participating in the research work-group DEPW, together with BouwhulpGroep, Cauberg Huygen, Delft University of Technology and the University of Maastricht. The illustrations in this paper are made by Ir. Arch. T.A. van de Straat.

Disclosure statement

No potential conflict of interest was reported by the author.

ORCID

E. E. Alders 🆔 http://orcid.org/0000-0002-8574-2032

References

Alders, E. E. 2016. "Adaptive Thermal Comfort Opportunities for Dwellings; Providing Thermal Comfort Only When and Where Needed in Dwellings in the Netherlands." *Doctoral thesis*, Delft University of Technology.

DGMR, Bouw BV. 2006. "Brochure Referentiewoningen Nieuwbouw." In edited by VROM. Sittard: SenterNovem.

Hoes, P. 2014. "Computational Performance Prediction of the Potential of Hybrid Adaptable Thermal Storage Concepts for Lightweight Low-energy Houses." *PhD*, Technical University of Eindhoven.

KNMI 2014. "Klimaatscenario's voor Nederland." In. Zwolle.

NEN_5060 2008. *NEN_5060+A2.xlsx+B2.xlsx+C2.xlsx; Hygrothermal performance of buildings - Climatic reference data*. Edited by Normcommissie 351 074 "Klimaatbeheersing in gebouwen". Delft: NNI.

NIWI 2002. "Tijdsbestedingsonderzoek 2000 TBO'2000." Netherlands Institute for Scientific Information Services.

Oldewurtel, Frauke, Alessandra Parisio, Colin N. Jones, Dimitrios Gyalistras, Markus Gwerder, Vanessa Stauch, Beat Lehmann, and Manfred Morari. 2012. "Use of Model Predictive Control and Weather Forecasts for Energy Efficient Building Climate Control." *Energy and Buildings* 45: 15–27. doi:10.1016/j.enbuild.2011.09.022.

Peeters, L., R. d Dear, J. Hensen, and W. D'Haeseleer. 2009. "Thermal Comfort in Residential Buildings: Comfort Values and Scales for Building Energy Simulation." *Applied Energy* 86 (5): 772–80. doi:10.1016/j.apenergy.2008. 07.011.

Thermal comfort and indoor air quality in super-insulated housing with natural and decentralized ventilation systems in the south of the UK

Paola Sassi ⓘ

ABSTRACT

Improved energy performance standards are resulting in better insulated and more airtight building. In such buildings, ventilation can be provided by natural means alone or in conjunction with extract mechanical ventilation or with whole-house mechanical ventilation with or without heat recovery. This paper reports on a study funded by the NHBC Foundation of the indoor environment of eight super-insulated homes with natural and decentralized ventilation systems in the south of the UK. The aim was to examine the effectiveness of such ventilation options. The buildings were monitored for one year in relation to temperature, relative humidity, CO_2, CO, NO_2, CH_2O and TVOC. The building occupants' feedback and IES building modelling triangulated the site data. The study showed that natural and decentralized ventilation systems provided good air quality in the case-study buildings and allowed users to create comfortable thermally differentiate environments in response to their preferences.

Introduction

Climate change and the drive for low-energy buildings have resulted in increasingly insulated and airtight buildings. In heating-dominated climates, the better insulated and airtight the buildings are the shorter the heating season and the less energy is needed to create comfortable homes in winter. In the UK, space and water heating account for approximately 80% of energy consumption (DBEIS 2016); therefore, reducing heating energy is critical to achieving the UK government goal of an 80% reduction in carbon dioxide (CO_2) emissions by 2050 from 1990 levels (UK Government 2008) and keeping global warming within the 2°C believed to mitigate risks, impacts and damages (Meinshausen et al. 2009).

Creating well-insulated and airtight buildings requires careful consideration of the provision of adequate ventilation to the building to ensure good air quality and thermal comfort. While studies about indoor pollutants and measured indoor temperatures in well-insulated and airtight buildings in the UK are limited; overheating in buildings has been the focus of a number of studies that have highlighted that even in mild maritime climates, such as that of the UK, overheating is already being experienced in buildings of different construction types, including energy-efficient and inefficient construction types (AECOM 2012; NHBC 2012; Mavrogianni et al. 2015; Zero Carbon Hub 2015). The overheating potential of buildings is going to increase as ambient temperatures rise with global warming. The Intergovernmental Panel on Climate Change's Fifth Assessment Report on Climate Change (Pachauri and Meyer 2014) predicts that ambient temperatures will rise and in the south of the UK

and this is expected to result in a 4°C increase of the mean summer temperatures and a 2–3°C increase of the mean winter temperature by the 2080s under a medium emissions scenario (Jenkins et al. 2009).

The ventilation system of a dwelling contributes significantly to its indoor air quality (IAQ) and the thermal comfort of its occupants. Domestic ventilation options include natural, mechanical, centralized and decentralized systems. Within the context of climate change, if adequate ventilation and thermal comfort can be provided by natural and decentralized ventilation systems, these would offer lower embodied energy and maintenance solutions compared to centralized mechanical systems (Beko, Clausen, and Weschler 2008). Furthermore, they have also been shown to be potentially associated with reduced operational energy (Sassi 2013). Decentralized systems can also represent less-disruptive and less-expensive solutions for high-performance retrofits of the existing housing stock, which is currently overwhelmingly naturally ventilated (Taylor et al. 2014), thus potentially facilitating the mainstreaming of such work.

This research aimed to identify any clear limitations of relying on natural and decentralized ventilations systems in well-insulated and airtight buildings in respect of IAQ and thermal comfort. The research also aimed to evaluate the operation of such buildings in relation to the occupants' interaction and their perception of comfort.

Eight highly insulted dwellings with decentralized ventilation systems were monitored, including the IAQ, the temperature and relative humidity. The research was funded by the NHBC Foundation.

Research method

Eight highly-insulated homes ventilated through decentralized and natural systems in the south of the UK were monitored for one year. The dwellings were chosen to provide a selection of different construction types, including heavy and light weight construction, and ventilation types, including systems based on the use of passive vents and through the wall mechanical extracts. Buildings' detailed plans and specifications were used to calculate the key parameters for comparing the buildings and assessing their performance. The dwellings that had not previously been tested for airtightness were tested. The building data were used to simulate the performance of the buildings in IES to simulate changes in occupancy, airtightness and ventilation and to allow for an additional level of comparison between the building's ventilation systems.

For a period of one year, measurements were taken for temperature and relative humidity at 30-minute intervals. Temperature loggers were placed in four rooms of each dwelling on different levels and with different orientations and including a living room and a bedroom. Relative humidity loggers were placed in the living room and one or two other rooms. The loggers used included the Hobo U10 and U12 (Temperature measurement range: $-20°C$ to $+70°C$, Relative humidity range: 25%(U10)/5%(U12) to 95%) and Tinytag Ultra temperature only and temperature and RH combined (Temperature measurement range: $-25 °C$ to $+85°C$, Relative humidity range: 0–95%). CO_2, CO, NO_2, CH_2O and TVOC measurements were taken over two-hour periods on three visits to the dwellings during different seasons. A Wolfsense IQ-604 probe was used with CO_2, CO, temperature and RH sensors installed plus an additional SEN-0-NO_2 Nitrogen Dioxide sensor and SEN-B-VOC-PPB Low range PID sensor b(0–20,000 ppb) for VOC's to take measurements every minute. A Formaldehyde meter (Wolfsense FM-801) was used to measure average levels over a period of an hour. Trend measurements of the indoor air pollutants were taken in one of the case-study buildings over several months in winter.

In addition, building occupants were interviewed in relation to their perceived comfort levels and their use of the building, including the adaptive actions taken to achieve comfort at three times throughout the year to gain feedback in respect of different seasons and weather conditions.

Ventilation strategies selection and effectiveness expectations

Air is introduced in buildings from outside through infiltration and ventilation and this dilutes pollutants in buildings, subject to the air outside being less polluted that that indoor. Infiltration is defined in the Building Regulations (2010, 13) Approved Document F1, Means of Ventilation as 'the uncontrolled air exchange between the inside and outside of a building through a wide range of air leakage paths in the building structure'. This is in contrast with ventilation that is controlled and provided through natural or mechanical means (Building Regulations 2010). The regulations differentiate between buildings with higher and lower infiltration rates and require different solutions for each. Buildings that are tested to have a higher infiltration rate than 5 m^3/hm^2 at 50 Pa are assumed to have air change rate per hour of 0.15 at ambient pressure, which will contribute to the fresh air provision in the building and consequently the area of controlled ventilation can be reduced compared to buildings with less air infiltration.

The case-study buildings all have decentralized and naturally ventilated systems with operable windows that provide purge ventilation as required. They fall into two of the four main types of ventilation set out in the Building Regulations ADF1 (2010), which include: trickle and other vents in conjunction with intermittent mechanical extract (five of the case studies can be classed as operating with such a system); passive stack ventilation system (three case studies use this system); continuous mechanical extract (centralized or decentralized) and continuous mechanical ventilation with heat recovery (MVHR).

Ventilation systems and differences in effectiveness

Three performance issues were examined in this study. The first relates to air quality. The provision of fresh air in relation to the volume of the building together with the control of sources of indoor air pollutants are the main influences on IAQ. The effectiveness of the natural ventilation that uses temperature differences and wind pressure to drive the ventilation through a passive stack system or windows is subject to the external weather conditions, obstructions, wind direction and speed, the internal and external building configuration and the design and use of windows and other openings. The quality of the indoor air can therefore vary and this research aims to establish whether in the case-study buildings the IAQ was sound despite such variations. Mechanical ventilation is independent of variables external to the building and only marginally affected by internal layouts (Clancy 2011) and mechanically ventilated buildings have been shown to benefit from good IAQ.

The second issue relates to winter thermal comfort and user preferences. Achieving thermal comfort in winter is as dependent on the heating system as the ventilation strategy. Decentralized systems tend to create thermal zones with different temperate within a building, while centralized systems tend to provide uniform temperatures in all rooms. The choice of a heating and ventilation system for winter performance can be more related to user preferences than to cost effectiveness of the system. The relationship between heating and ventilation systems and their energy use was not investigated in this research due to the extensive use of timber wood stoves in the case-study buildings.

The third and last point relates to the effectiveness of natural ventilation in achieving summer thermal comfort. Summer thermal comfort in UK homes is predominantly achieved through the opening of windows even in buildings with centralized ventilation systems. While overheating has been recorded in poorly insulated buildings as well as in well-insulated buildings, including certified Passivhaus dwellings (AECOM 2012; Mcleod, Hopfe, and Kwan 2013; Mavrogianni et al. 2015), increased insulation of buildings results in the retention of internal and solar gains within the building, potentially creating uncomfortable environments. If ambient temperatures are above the comfort level, then exterior air cannot be used to cool interior environments. At present the ambient temperatures are only seldom above comfort levels and therefore appropriate for providing direct cooling of occupants, subject to the configuration of the building and the ventilation openings providing effective air changes. The effectiveness of the individual ventilation systems and their design was of particular interest in the case-study buildings.

IAQ and measurement results

Contaminants of indoor air in buildings can include human bio-effluents (including carbon dioxide (CO_2)), external pollution from vehicles, volatile organic compounds (VOCs) (including formaldehyde (CH_2O)), tobacco smoke, radon, ozone, carbon monoxide (CO), oxides of nitrogen (including nitrogen dioxide (NO_2)), bacteria, fungal spores, mites and fibres (ISO 2008).

Airtight construction in conjunction with natural ventilation, which is not automatically controlled, can result in reduced air changes and heighten the risk of accumulation of pollutants and CO_2. In addition to the infiltration rates (which in the buildings analysed varied between 0.4 and 7 air changes per hour at 50 Pascals), the concentration of pollutants and CO_2 is also related to the volume of air in the building within which the pollutants can diffuse and the occupation density (which in the buildings analysed ranged from 66 m^3 of air per person to 240 m^3 per person).

In all case-study buildings, occupants were conscious of using consumer products that had low VOCs and only using those they felt really necessary, for instance none of the occupants used air fresheners. Most building materials were typically low-emissions options such as timber rather than carpet flooring.

Indoor pollutants can have minor to severe impacts on occupant's health, which, depending on the susceptibility of the occupants and their level of exposure, can include sensory irritation, causing fatigue, headache and shortness of breath, chronic pulmonary disease, cancer and death (Chianga and Laib 2002; Daisey, Angell, and Apte 2003; Kephalopoulos, Koistinen, and Kotzias 2006; WHO 2010; Clancy 2011).

Of the 'classical' pollutants monitored (CO_2, CO, NO_2, CH_2O and TVOC), as defined by the Scientific Committee on Health and Environmental Risks (SCHER 2007), CO, CH_2O and NO_2 are classified as high-priority chemicals in the European Commission publication 'Critical appraisal of the setting and implementation of indoor exposure limits in the EU' (Kotzias et al. 2004).

Carbon monoxide (CO) and formaldehyde CH_2O

CO poisoning is a leading cause of death from indoor chemical pollution (Kotzias et al. 2004; WHO 2010). CO is produced as a result of incomplete combustion of fuels in faulty, poorly maintained or ventilated cooking and boiler appliances, or open fires burning biomass fuel. Tobacco smoke also is a source of CO (Kotzias et al. 2004). CH_2O is a known animal and human carcinogen and even at low concentrations, lower than those associated with cancer, it can cause sensory irritation (WHO 2010). Building and furniture board materials are a source of CH_2O as is tobacco smoke. All monitored buildings had low levels of CO and CH_2O and the results' confidence was high (Table 1).

Nitrogen dioxide (NO_2)

NO_2 results from the burning of fossil fuel both indoors (cooking and heating appliances) and outdoors (motor vehicles). Elevated levels in relation to the German indoor guidance level of 60 µg/m3 (31 ppb) were found in 25% and 45% of dwellings in Germany and Italy, respectively (Kotzias et al. 2004). WHO identified research suggesting NO_2 being linked to an impairment of bronchial function, including research that linked a 20% increased risk of lower respiratory illness in children exposed to elevated NO_2 levels from 15 µg/m3 (8 ppb) to 43 µg/m3 (23 ppb) (WHO 2010). The monitored equipment used in this research lacked adequate sensitivity to provide high confidence in the results taken over a period of typically only one hour. However, some elevated levels were noted in case studies 2, 3 and 8, with potential sources being external traffic, cooking and wood-burning appliances, respectively. It is worth noting that in addition to Case Study 8 also case studies 1, 4, 5 and 7 had wood-burning stoves but elevated levels of NO_2 were not noted.

Additional experiments in two non-insulated and non-airtight homes confirmed that cooking with gas and wood-burning stoves are significant sources of NO_2. The first experiment measured the NO_2 emissions from a wood-burning stove. Figure 1 shows a temperature rise from 10°C to 20°C when the woodburner was lit. NO_2 is formed at high temperatures; therefore for the first hour NO_2 levels in the living room are around zero. As the woodburner reaches sufficiently high temperatures, NO_2 is formed and NO_2 levels in the living room reach 60–70 ppb. Even after six hours NO_2 levels are still above recommended long-term exposure levels.

The second experiment simulated a 25-minute cooking process using two gas burners. Five ventilation options were tested and the levels of NO_2, CO_2, CO and TVOC were measured. CO and TVOC were not of concern, but NO_2 and CO_2 levels peaked at

Table 1. Winter measurements of IAQ taken in the eight case-study buildings, over a period of 90 min average and selected exposure standards.

Chemical	CO ppm	CH_2O ppb	NO_2 ppb	TVOC µg/m^3	CO_2 ppm
Compulsory standards and exposure limits and WHO (2010) standard	90 ppm – 15 mins; 50 ppm – 30 mins; 25 ppm – 1 hour; 10 ppm – 8 hour (Building Regs F1 2010)	80 ppb over a 30-min period and long-term exposure (WHO 2010)	150 ppb – 1 hour; 20 ppb – long-term exposure (Building Regulations F1 2010)	300 µg/m^3 (Building Regulations F1 2010)	School average levels for full day not to exceed 1500 ppm (Building Bulletin 2006)
Voluntary Well Building Standard (Delos Living LLC 2015)	9 ppm	27 ppb		500 µg/m^3	800 ppm
Case Study 1	1.2	10–15	0–49	257	838
Case Study 2	1.6	10–15	12–55	307	1224
Case Study 3	0.1	10	35–75	66	706
Case Study 4	1.7	25	0–45	289	747
Case Study 5	1.1	10	0–39	307	691
Case Study 6	4.3	10–20	0–47	573	1086
Case Study 7	0.4	11	0–44	297	735
Case Study 8	0.1	20–29	16–52	205	1087

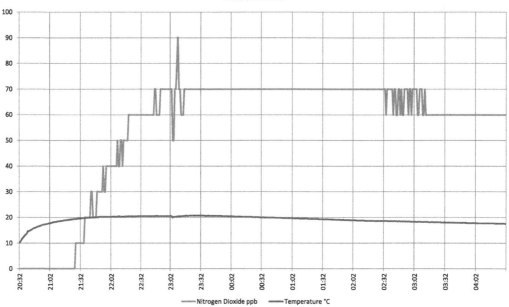

Figure 1. Nitrogen dioxide emissions from wood-burning stove.

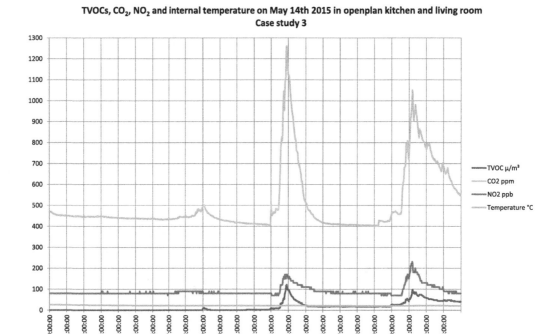

Figure 2. Nitrogen dioxide emissions from cooking in Case Study 3.

356 ppb and over 4000 ppb respectively in the poorly ventilated options tested and took half an hour to drop back to normal levels. Opening internal or external doors and windows as well as using the extract hood proved effective in keeping all chemical levels below those of concern. The impact of gas cooking could also be clearly seen in Case Study 3's open plan kitchen living room (Figure 2). NO_2, CO_2 and TVOCs rise in line with cooking activities on a gas hob. While TVOCs and CO_2 peak below the levels of concern, the levels of NO_2 are briefly above the levels of concern set by the WHO.

Total volatile organic compounds (TVOCs)

TVOC is a measure of combined volatile organic compounds. These include chemicals such as benzene, toluene and tetrachloroethylene and other carbon-based chemicals. Sources of VOCs in buildings include materials and furniture, leather and textiles, paints, varnishes, sealants, thinners, adhesives, household products (cleaning products, pesticides, moth repellents, air fresheners) and personal care products (cosmetics, perfumes) (European Commission 2002). VOCs are differentiated according to their boiling points and classified as VVOC (very volatile organic compounds); VOC (volatile organic compounds); SVOC (semi-volatile organic compounds). Background levels are around 0.05–0.4 ppm (Wolfsense 2014). According to research by Kephalopoulos, Koistinen, and Kotzias (2006), more than 900 VOC have been identified in buildings, 250 have been measured at concentrations higher than 1 ppm and typically in one building VOC levels are usually lower than 1–3 mg/m³. The health impacts are primarily of a sensory nature. Recommended exposure levels are difficult to formulate due to the mixture of chemicals and measuring techniques and WHO does not state any recommended exposure limits. Research attempting to define exposure levels has derived exposure levels from sensory responses or from statistical surveys of existing levels (Seifert 1999). The TVOC levels measured in the case-study buildings were all within The Well Building Standard of 500 µg/m³ and of a high confidence level. The highest levels were measured in Case Study 6 where the occupants smoke indoors (446 µg/m³), and these exceed the Building Regulations (2010) standard of 300 µg/m³. Slightly elevated measurements were noted and ascribed to the use of craft and similar products associated with leisure activities.

Carbon dioxide CO_2

CO_2 is not considered a health hazard in its own right (ISO 2008). Extremely high levels above 10,000 ppm of CO_2, which are not normally found in buildings, can cause drowsiness and at much higher levels can cause unconsciousness (Clancy 2011). However, lower levels that can be found in buildings, such as 1000–2500 ppm, have been found to moderately (1000 ppm) to significantly (2500 ppm) detrimentally affect approximately two-thirds of specific decision-making office-based activities (Satish et al. 2012) but have no physiological impact.

Occupants are the main source of CO_2 as well as other bio-effluents (such as body odour) that might be unacceptable to other occupants (Dougan and Damiano 2004; Petty n.d.). Being linked to occupancy, particularly in commercial buildings, CO_2 has been used as an indicator of ventilation rates and used as a basis for designing ventilation solutions; however levels of CO_2 are not necessarily directly linked to levels of other pollutants (Dougan and Damiano 2004; Nga et al. 2011). The measurements taken in the case-study buildings illustrate this point. Figure 3 shows levels of CO_2 rise with occupancy while the TVOC levels slightly decrease, indicating the two are not linked.

The CO_2 levels measured in half the case studies were within The Well Building Standard (Delos Living LLC 2015) limit of 800 ppm and half above that but within the Building Bulletin (2006) target of 1500 ppm. Despite the level being above the 800 ppm, the occupants who rated their environment on a seven-point Likert scale perceive their environment as being fresh and not stuff or smelly.

Conclusion related to IAQ

While the small sample of case studies precludes any generalizations, this research has not shown any reasons in relation to IAQ to discourage the use of natural and decentralized ventilation systems in airtight and well-insulated housing.

Thermal comfort and measurements results

To help define acceptable indoor temperatures, the following standards and guidance were considered: ASHRAE Standard 55 (2010), the British Standard (2007) BS EN15251:2007 and CIBSE Guide A: Environmental Design (2015).

For summer comfort, according to BS EN 15251:2007 the acceptable internal temperatures would rise with the external

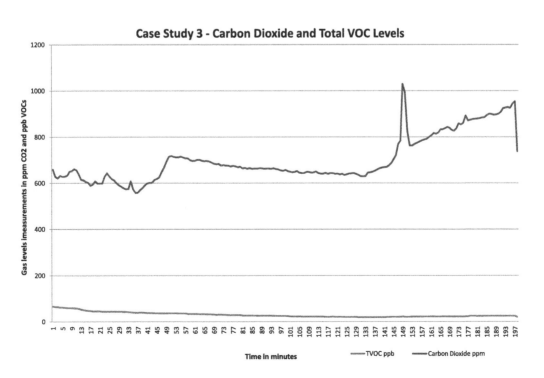

Figure 3. Example of relation of CO_2 to TVOC levels.

temperatures in line with the adaptive thermal comfort model. The formula to calculate the indoor maximum compared with external temperature is:

indoor maximum (T_{max}) = 0.33 *external running mean temperature (T_{rm}) + 18.8 + 2, +3, or +4 depending on the category of building being monitored. This building should be classed at 'category 2' (normal expectation should be used for new buildings and renovations) for which the equation for maximum recommended temperature is given by

$$T_{max} = 0.33 * T_{rm} + 21.8.$$

Note that T_{rm} is the running mean of the outdoor temperature, not the daily maximum

This would mean that an external running mean temperature of 20°C (fairly normal in UK summer conditions) would result in a maximum internal temperature of 21.8 + 6.6 = 28.4°C to feel acceptable for occupants. For conditions such as those shown in Figure 7 with a T_{rm} of about 16°C the maximum acceptable temperature will be about 27°C. This model acknowledges that human adaptation through clothing but also a physiological adaptation can raise the maximum temperature considered comfortable by most people in summer.

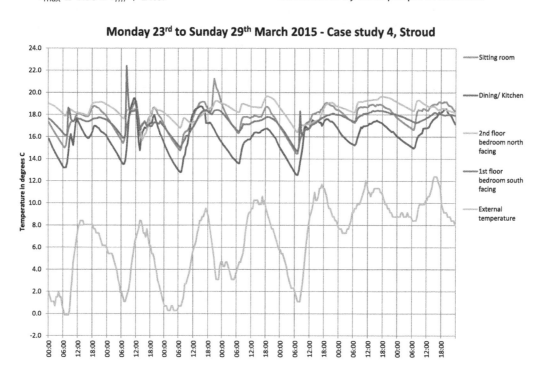

Figure 4. Typical winter performance of case-study building showing cool to cold kitchen area.

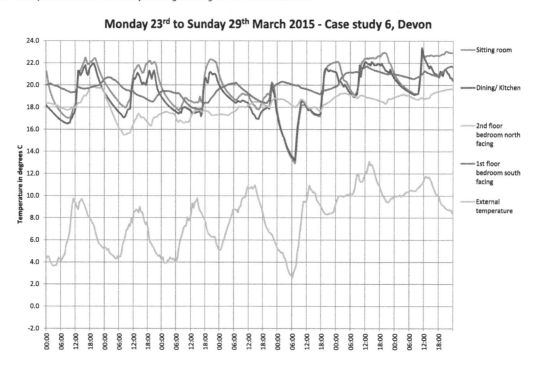

Figure 5. The living and kitchen rooms are between 2°C and 4°C warmer than the bedrooms.

For winter comfort, the BS EN 15251:2007 standard recommends 18–21°C in living spaces, including bedrooms, and 14–18°C for other spaces such as storage and halls. These limits probably apply in the conditions shown in Figures 4 and 5, where the mean external temperature is below 10°C. CIBSE Guide A: Environmental Design suggests a wider range of temperatures for different rooms and seasons (Table 2).

The indoor comfort temperature set by CIBSE Guide A (2015) for the summer are 25°C for living rooms and 23°C for bedrooms. Overheating is deemed to have occurred if three percent of the occupied hours over one year exceed the recommended maximum indoor temperature T_{max}. CIBSE Guide A also notes that high bedroom temperatures over 24°C can impair sleeping and this suggests that it is important to differentiate when the peak temperatures occur.

The thermal experience in well-insulated and airtight buildings: the principles

Air and radiant temperature, air movement, humidity and the user activity, clothing habits and ability to control their environment affect the user perception of thermal comfort (Nicol, Humphreys, and Roaf 2012). Winter and summer comfort parameters are discussed below.

Regardless of the ventilation strategy, a highly insulated building fabric creates internal surfaces that are warmer than in poorly insulated buildings. The human body's perception of thermal comfort is affected significantly by the radiant heat exchange with surrounding surfaces. If those surfaces are cold, like that of a single glazed window, the air temperature needs to be suitably high to offset the heat loss experienced through radiant heat exchange with the cold surface. If the internal building surfaces are warm, the internal air temperature can be lower and still provide a comfortable environment for the occupants. In relation to air movement, in airtight buildings, users tend to experience draughts less.

The ventilation strategy can affect the winter comfort experience due to its impact on the humidity, temperature distribution and the perceived control of the users to influence their environment. The typically higher ventilation rates experienced in mechanically ventilated buildings can cause low relative humidity levels, which when below 30% can cause dryness to the skin and mucus membranes, potentially increasing the vulnerability of throat and nose to viruses. Conversely where ventilation rates are low the relative humidity levels can rise above recommended levels and while this may not be perceived by users, it can cause mould growth, notably where cold bridges occur in the construction. The perceived personal control of the internal temperature can be missing in centrally mechanically ventilated buildings, and even though control is typically available

it may appear complex and unusual. In contrast in naturally ventilated buildings, the occupants have the benefit of being able to manipulate their building to make it more comfortable and this control facility is also known to make occupants more tolerant of their environment (Baker and Standeven 1996; Brager and de Dear 1998). The temperature distribution is affected by the system of heat provision in the building and in mechanically ventilated homes, especially those with heat recovery and top-up heating integrated within the ventilation systems, the temperature tends to be constant throughout the whole building. Decentralized ventilation systems in conjunction with room controlled heating sources can more readily create different thermal zones within a building that respond to different users' preferences.

The summer comfort within a building depends on the building's ability to exclude solar gains with insulation and shading, draw in and move cooler air within the building and expel hot air from the building, and absorb excessive heat within the building fabric to avoid raising the internal temperature (Porritt, Shao et al. 2011; Porritt, Cropper et al. 2012; Mavrogianni et al. 2014). The choice of mechanical centralized ventilation is particularly aimed at achieving good winter performance and in summer such buildings often also use the opening of windows as a means of cooling the space. The use of MVHR per se has not been shown to avoid overheating, and buildings with MVHR, including certified Passivhaus dwellings, have been shown to overheat in southern, central and northern Europe (Mcleod, Hopfe, and Kwan 2013).

The thermal experience in well-insulated and airtight buildings: the experience in the case-study buildings in winter

The winter experience in the case-study buildings reflected the impact of the radiant surface temperature on the perception of thermal comfort. All occupants reported temperatures between 17°C and 22°C being comfortable, suggesting that the impact on higher radiant surfaces could have improved the perceived comfort at lower temperatures. Three buildings experienced lower than 17°C temperatures, two of which were the same building design that included a lower level kitchen, which the users considered cool to cold at between 15°C and generally not more than 18°C (Figure 4). The third building experienced temperatures of 15–16°C at times in the living room with an average temperate of 17°C during daytime hours in winter, but the users considered the environment comfortable. The latter example may well have combined the physiological impact of radiant temperatures with the psychological impact of having control over the environment and indeed having designed the space, as the owner is also the building designer.

The user feedback also reflected the fact that occupants in less airtight buildings can experience more air movement. However the case-study occupants who noted the air movement did not consider that uncomfortable and did not express a wish to reduce the air movement.

User preferences also clearly manifested themselves in the monitoring data. The thermal requirements of different rooms in a house depend on how the building is used, whether it is used by a family or single persons or as shared house. For instance,

Table 2. Range of internal temperatures in °C considered appropriate for different rooms in dwellings during winter and summer from CIBSE Guide A (2015).

	Winter	Summer
Bathrooms	26–27	26–27
Bedrooms	17–19	23–25
Hall/stairs/landings	19–24	21–25
Kitchen	17–19	21–23
Living rooms	22–23	23–25
Toilets	19–21	21–23

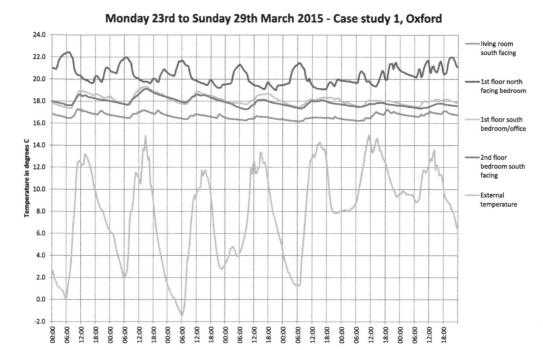

Figure 6. In this shared house, each occupants heat their personal space to their preferred temperature. The living room is less used by the occupants and not heated as much, not only because it is used infrequently but also because it is open plan and linked to a dining area and therefore constitutes a large and difficult to heat space.

Case Study 6 is used as a family home and Figure 5 shows how the kitchen and living room are heated to 21–22°C, while the bedrooms are between 16°C and 20°C. Case Study 1 (Figure 6) is a shared house and the occupants choose to heat their own rooms according to their personal preferences. Both Case Studies 1 and 6 successfully provide the opportunity to create different thermal zones that suit a variety of potential users. Most traditional homes with cellular arrangements of spaces offer the opportunity to create separate thermal zones.

The thermal experience in well-insulated and airtight buildings: the experience in the case-study buildings in summer

The summer performance of the case-study buildings relied, like most dwelling in the UK, on opening windows to provide ventilation. One of the case studies only had shading on the whole south-facing façade; the other cases studies had limited shading provided by curtains. Three of the case studies had thermally massive floor and walls with timber framed roof; the others had a thermally massive ground floor and timber framed walls and roof. The case studies monitoring highlighted a number of phenomena that are instructive when considering the design for summer comfort through natural ventilation. Some confirm or question well-understood principles; others relate to less common considerations.

Rooflights

Avoiding solar gains is a fundamental aim of creating a thermally comfortable internal environment in summer. During the summer months when the sun is at its highest point in the sky, the most exposed glazed surfaces are those that are horizontal. Therefore at the hottest time of the year, rooflights provide

significant unwanted heat gains. Data from Case Study 3 illustrated this in relation to the small bedroom on the second floor, which has a large rooflight. On one of the hottest days of 2016 the building was unoccupied and sealed. The internal temperatures in second floor bedroom rose from 25°C to 40°C in 10 h before the occupant returned and opened the rooflight, allowing in external air, which at 34°C was still very hot but cooler than the internal air. It is important to note that all other spaces with south-facing vertical glazing were 6°C cooler (peaking at 34°C). If rooflights are needed, providing external shading or roller shutters that can prevent the solar gains to enter the building is essential.

Thermal mass

The data showed no direct relationship between overheating in lightweight compared to heavy weight construction. While all the case studies were different in design and context and it would have been difficult to assess through monitoring the impact of thermal mass, the results suggest that through appropriate design a comfortable environment can be achieved in the current UK climate with light weight construction. The modelling of the case studies in IES confirmed only minimal impact of thermal mass in the case-study buildings.

Ventilation design

To provide effective cross or single-sided ventilation, stack ventilation or night time ventilation, the window design and the configuration of openings is critical. The case studies included examples of effective cross and stack ventilation.

The effectiveness of cross ventilation can be seen in Case Study 2 bedroom on the second floor. Figure 7 shows how upon return to the house at 18.30, the occupant of the bedroom opens the window and rooflights on either side of the room (Figures 8 and 9) and the temperature drops 3°C in three hours. At just

14th April 2015 - Case study 2, Oxford

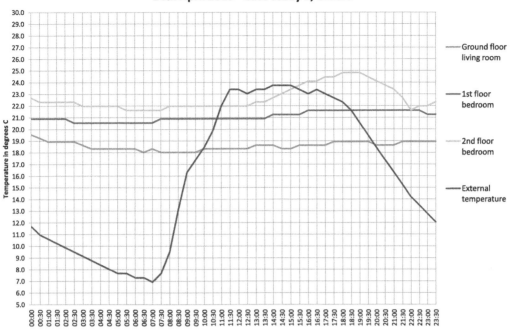

Figure 7. Case Study 2 temperature experienced in three locations within the building on a typical warm to hot day.

Figure 8. Case Study 2 top floor bedroom has good cross-ventilation. View of the rooflight on one side of the room.

Figure 9. Case Study 2 view of the window on south-west side of the room to provide cross-ventilation.

Figure 10. Louvered window proving safe ventilation.

above 21°C the window is closed as the temperature is considered comfortable. It is also worth noting that the north-east facing bedroom on the 1st floor with no direct solar gain and very well insulated in all directions (floor, roof and walls) retains a stable temperature of between 20.5°C and 21.5°C through the whole day.

Case studies 4 and 5 are examples of effective stack ventilation in practice. Both have the same design that includes a large

Table 3. Temperatures in °C monitored during summer heat wave June–July 2015 (minimum and maximum temperatures are shown in parentheses).

Case studies M = masonry T = timber frame MV = vents and decentralized extracts PV = Passive vent system	Living room south ground floor	Living room west	Kitchen north	Living room north ground floor	Bedroom east ground floor	Living room south north facing	Room 1st fl. south facing	Bedroom 1st fl. north facing	Bedroom 2nd fl. south facing	Bedroom 2nd fl. north facing	External temperature
1 – Oxfordshire – M – MV	22.02 (20.5) (23.4)						23.28 (20.9) (27.4)	23.37 (22.0) (25.9)	25.12 (21.6) (29.2)		18.68 (11) (32.2)
2 – Oxfordshire – M – MV	21.72 (19.9) (23.7)								24.20 (19.8) (32.2)		18.68 (11) (32.2)
3 – London – T – MV	25.82 (21.3) (33.7)					24.45 (19.5) (33.7)	25.94 (21.3) (34.1)		26.24 (19.8) (39.8)		20.93 (11.3) (38.5)
4 – Gloucestershire – T – MV	21.09 (15.0) (30.0)			20.80 (14.5) (29.4)			21.33 (15.5) (29.7)			20.85 (14.3) (30.2)	17.94 (10.2) (31.1)
5 – Gloucestershire – T – MV[a]	23.03 (20.0) (27.5)			23.11 (20.2) (27.7)			25.16 (21.9) (30.0)			25.04 (22.0) (29.7)	17.94 (10.2) (31.1)
6 – Somerset – T – PV[b]	20.81 (18.4) (19.2)		20.25 (19.1) (20.1)				21.10 (18.6) (21.7)	22.54 (21.3) (22.4)			17.65 (12.2) (27.1)
7 – Somerset – M – PV[b]	23.37 (21.9) (24.9)				23.94 (23.3) (24.5)		22.46 (21.5) (23.4)	23.39 (22.3) (28.7)			17.65 (12.2) (27.1)
8 – Somerset – M – PV[b]	19.44 (19.1) (20.1)	18.73 (18.4) (19.1)					21.92 (21.3) (22.4)	20.43 (18.6) (21.7)			17.65 (12.2) (27.1)

[a]Occupants on holiday over a two-week monitoring period of heat wave.
[b]Monitoring period (24th–27th June) did not include peak heat wave.

Table 4. Distribution of temperatures in °C measured in living room in case study 4 as percentage of overall hours over heat wave period.

15°C	16°C	17°C	18°C	19°C	20°C	21°C	22°C	23°C	24°C	25°C	26°C	27°C	28°C	29°C	30°C
1.2%	1.5%	3.6%	8.5%	20.4%	22.2%	15.3%	10.9%	7.6%	2.1%	2.0%	1.4%	1.2%	1.1%	0.9%	0.2%

stair case that is open to the lower two split levels. This config-uration allows air to move freely and creates an effective stack effect where hot air can exit through a rooflight at the top of the stairs. Case Study 1 also has a central stairs designed to draw air up, but the stair small and enclosed and the lack of connec-tions to the living spaces around it means it is not perceived by the occupants as working effectively as a means of passive stack ventilation.

Safety is an increasing concern and it was noted that several of the occupants did not open the windows when they were outside the home for safety reasons. Safe ventilation options are available to address such concerns. The Hanson EcoHouse at the BRE Innovation Park includes windows with safe night ven-tilation louvers (Figure 10), which may need to become more commonly used.

Thermal zones

The performance during a two-week period (which included a heat wave) demonstrated a clear distribution of heat in all the case-study buildings, which was primarily vertical. Case Study 1 ground floor was cooler than the first floor, which was cooler than the second floor. Similar stratification can be seen in the other case studies in Table 3 regardless of whether includ-ing heavy or lightweight construction. To a lesser degree the south-facing spaces were warmer than the north-facing ones. Some apparent anomalies such as Case Study 7 north-facing bedroom being hotter than the south-facing bedroom can be explained by the existence of a large rooflight in the north-facing room.

Table 3 also shows the peak temperatures reached during a two-week hot period and it is worth noting that virtually all case studies performed reasonably well in relation to the CIBSE Guide A (2015). For instance, Case Study 4, which experienced the high-est peak temperatures outside London, experienced tempera-tures over 25°C in the living room for only a small percentage of hours (4.7%) (Table 4). In the bedrooms over the two-week period the temperature exceeded 24°C between 22.00 and 8.00 for 7.5 h of which 4 h were below 25°C. Also important to note is that the occupants overall felt comfortable in their homes, even if they judged the environment to be slightly warm or even too warm.

Discussion and conclusion

The case studies monitored were sufficiently varied that a direct comparison between case studies is not appropriate but some observations can contribute to a better understanding and design of ventilation system and buildings.

IAQ

(1) Overall the study suggests that decentralized ventilation systems in highly insulated buildings can provide adequate to good IAQ. While IAQ measurements taken in summer time, when the windows were mainly kept open, were bet-ter than in winter, the pollutants' levels measured in winter do not suggest unhealthy environments.

(2) While the data related to the case-study buildings were unclear, additional tests showed that any burning of fossil fuels can be associated with NO_2 emissions and that while these are short-lived and local their avoidance would be preferable.

(3) The data also confirm other studies that found no direct relation between CO_2 levels and levels of TVOC or other chemicals. In the case studies investigated, the CO_2 levels in some case studies were above the Well Building Standard recommendations of 800 ppm; however the perception of the occupants was still of good air quality.

Thermal comfort

(1) The occupants' surveys reflected a high level of thermal satisfaction experienced by all occupants. Even when the spaces were considered slightly too warm or slightly too cold, the satisfaction was reported as high. Considering that the internal temperatures were often not in line with what the thermal comfort standards recommend, it is worth con-sidering that the standards may not yet reflect the wide range of thermal preferences experienced in reality.

(2) The user satisfaction cannot be considered without acknowl-edging the 'forgiveness factor' of failings experienced by users of spaces they are emotionally attached to, such the overheating experienced in Case Study 3. Also, having emo-tionally invested in a building meant in relation to the case studies that the occupants mainly, but not without excep-tion, knew how to best manipulate their home to create a comfortable environment.

(3) The case-study data identified some examples of good and effective practice, such as effective cross and stack venti-lation, but an equal number of ineffective solutions, which suggests a lack of adequate knowledge in the industry.

(4) The study also highlighted that some solutions, which would help us to create more comfortable environments especially in a scenario of climate change, are still underused in the industry, for instance external shutters and shading and secure ventilation openings.

(5) The thermal zones identified in the study also appear to be underappreciated by building designers and could be con-sidered in the design of buildings with natural and decen-tralized ventilation systems.

In conclusion, the study supports the potential for the use of decentralized and natural ventilation in housing in a mild climate, such as that of the southern UK. However, the study also shows that while the case-study buildings overall work well, some of the ventilation and thermal comfort solutions applied in the case-study buildings could be improved. Considering that architects were involved in the design of all case-study buildings, and keeping in mind the other reports of building failures related to well-insulated buildings (AECOM 2012; NHBC 2012; Mavro-gianni et al. 2015; Zero Carbon Hub 2015), one could conclude that it is essential for the building industry to achieve a better understanding of the operation of well-insulated buildings and in particular their ventilation, if the industry is to provide the enhanced energy building performance required to reduce the risk of climate change. To support such improvements, building professionals need to be aware of the significance of good ven-tilation design on the performance and perceptual success of designs. More research work is also required as well as the development of effective vehicles for disseminating the good practice in the field.

Disclosure statement

No potential conflict of interest was reported by the author.

Funding

This work was supported by NHBC Foundation.

ORCID

Paola Sassi ⓘ http://orcid.org/0000-0002-8422-0988

References

AECOM. 2012. *Investigation Into Overheating in Homes: Literature Review*. London: Department for Communities and Local Government.

ASHRAE (American Society of Heating, Refrigerating and Air-Conditioning Engineers). 2010. *Standard 55 Thermal Environment Conditions for Human Occupancy*. Atlanta, GA: ASHRAE.

Baker, N. V., and M. A. Standeven. 1996. "Thermal Comfort in Free-running Buildings." *Energy and Buildings* 23: 175–182.

Beko, G., G. Clausen, and C. Weschler. 2008. "Is the Use of Particle Air Filtration Justified? Costs and Benefits of Filtration with Regard to Health Effects, Building Cleaning and Occupant Productivity." *Building and Environment* 43: 1647–1657.

Brager, G. S., and R. J. de Dear. 1998. "Thermal Adaptation in the Built Environment: A Literature Review." *Energy and Buildings* 27: 83–96.

BSI (British Standards Institution). (2007). *BS EN 15251: 2007, Indoor Environmental Input Parameters for Design and Assessment of Energy Performance of Buildings Addressing Indoor Air Quality, Thermal Environment, Lighting and Acoustics*. London: BSI.

Building Bulletin. 2006. *Building Bulletin 101. Ventilation of School Buildings. Regulations Standards Design Guidance*. Version 1.4 – 5th July 2006.

Building Regulations. 2010. *Building Regulations Approved Document F1, Means of Ventilation*. London: NBS.

Chianga, C., and C. Laib. 2002. "A Study on the Comprehensive Indicator of Indoor Environment Assessment for Occupants' Health in Taiwan." *Building and Environment* 37, 387–392.

CIBSE. 2015. *CIBSE Guide A: Environmental Design*. London: Chartered Institution of Building Services Engineers.

Clancy, E. 2011. *Indoor Air Quality and Ventilation. CIBSE Knowledge Series: KS17*. London: The Chartered Institution of Building Services Engineers.

Daisey, J. M., W. J. Angell, and M. G. Apte. 2003. "Indoor Air Quality Ventilation and Health Symptoms in Schools: An Analysis of Existing Information." *Indoor Air* 13 (1): 53–64.

Delos Living LLC. 2015. *The Well Building Standard*. New York: International Well Building Institute.

Department of Business, Energy and Industrial Strategy (DBEIS). 2016. *Energy Consumption in the UK*. London: DBEIS.

Dougan, D. S., and L. Damiano. 2004. CO_2-based Demand Control Ventilation. Do Risks Outweigh Potential Rewards? *ASHRAE Journal*, October 2004, 47–53.

European Commission. 2002. *Screening Study to Identify Reductions in VOC Emissions Due to the Restrictions in the VOC Content of Products*. Final Report. February 2002, Brussels: European Commission.

ISO (International Standards). 2008. *Building Environment Design — Indoor Air Quality — Methods of Expressing the Quality of Indoor Air for Human Occupancy ISO 16814:2008(E)*. Geneva: International Standards.

Jenkins, G. J., J. M. Murphy, D. M. H. Sexton, J. A. Lowe, P. Jones, and C. G. Kilsby. 2009. *UK Climate Projections: Briefing Report*. Exeter, UK: Met Office Hadley Centre, United Kingdom Climate Impacts Programme.

Kephalopoulos, S., K. Koistinen, and D. Kotzias. 2006. *Strategies to Determine and Control the Contributions of Indoor Air Pollution to Total Inhalation Exposure (STRATEX)*. European Collaborative Action Report No 25. Luxembourg: European Commission.

Kotzias, D., K. Koistinen, C. Schlitt, P. Carrer, M. Maroni, and M. Jantunen. 2004. *Summary of Recommendations and Management Options. Critical Appraisal of the Setting and Implementation of Indoor Exposure Limits in the EU. EUR 21590 EN. INDEX Project*. Ispra, Italy: European Commission, Joint Research Centre.

Mavrogianni, A., M. Davies, J. Taylor, Z. Chalabi, P. Biddulph, E. Oikonomou, and B. Jones. 2014. "The Impact of Occupancy Patterns, Occupant-controlled Ventilation and Shading on Indoor Overheating Risk in Domestic Environments." *Building and Environment*, 78, 183–198.

Mavrogianni, A., J. Taylor, M. Davies, C. Thoua, and J. Kolm-Murray. 2015. "Urban Social Housing Resilience to Excess Summer Heat." *Building Research & Information* 43 (3): 316–333.

Mcleod, R. S., C. J. Hopfe, and A. Kwan. 2013. "An Investigation Into Future Performance and Overheating Risks in Passivhaus Dwellings." *Building and Environment* 70 (21): 189–209.

Meinshausen, Malte, Nicolai Meinshausen, William Hare, Sarah C. B. Raper, Katja Frieler, Reto Knutti, David J. Frame, and Myles R. Allen. 2009. Greenhouse-Gas Emission Targets for Limiting Global Warming to 2°C. *Nature* 458, 1158–1162.

Nga, M. O., M. Qua, P. Zhen, Z. Li, and Y. Hanga. 2011. "CO_2-based Demand Controlled Ventilation Under New ASHRAE Standard 62.1-2010: A Case Study for a Gymnasium of an Elementary School at West Lafayette, Indiana." *Energy and Buildings* 43, 3216–3225.

NHBC. 2012. *Overheating in New Homes. A Review of the Evidence. NF 46*. Milton Keynes: NHBC Foundation.

Nicol, F., M. Humphreys, and S. Roaf. 2012. *Adaptive Thermal Comfort. Principles and Practice*. London: Routledge.

Pachauri, R. K., and L. A. Meyer, eds. 2014. *Climate Change 2014: Synthesis Report. Contribution of Working Groups I, II and III to the Fifth Assessment Report of the Intergovernmental Panel on Climate Change*. Geneva, Switzerland: IPCC.

Petty, S. n.d. "Summary of ASHRAE'S Position on Carbon Dioxide (CO2) Levels in Spaces, Energy & Environmental Solutions, Inc." Accessed 26 January 2016. www.eesinc.cc/downloads/CO2positionpaper.pdf.

Porritt, S. M., P. C. Cropper, L. Shao, and C. I. Goodier. 2012. "Ranking of Interventions to Reduce Dwelling Overheating During Heat Waves." *Energy and Buildings* 55: 16–27.

Porritt, S. M., L. Shao, P. C. Cropper, and C. I. Goodier. 2011. "Adapting Dwellings for Heat Waves." *Sustainable Cities and Society* 1: 81–90.

Sassi, P. 2013. "A Natural Ventilation Alternative to the Passivhaus Standard for a Mild Maritime Climate." *Buildings* 3: 61–78.

Satish, U., M. J. Mendell, K. Shekhar, T. Hotchi, D. Sullivan, S. Streufert, and W. J. Fisk. 2012. "Is CO_2 an Indoor Pollutant? Direct Effects of Low-to-Moderate CO_2 Concentrations on Human Decision-making Performance." *Environmental Health Perspectives* 120 (12): 1671–1677.

SCHER (Scientific Committee on Health and Environmental Risks). 2007. *Opinion on Risk Assessment on Indoor Air Quality*, 29 May 2007. Brussels: European Commission.

Seifert, B. 1999. Richtwerte für die Innenraumluft. Die Beurteilung der Innenraumluftqualität mit Hilfe der Summe der flüchtigen organischen Verbindungen (TVOC-Wert). *Bundesgesundheitsblatt, Gesundheitsforsch – Gesundheitsschutz* 42, 270–278.

Taylor, J., C. Shrubsole, M. Davies, P. Biddulph, P. Das, I. Hamilton, and E. Oikonomou. 2014. "The Modifying Effect of the Building Envelope on Population Exposure to PM2.5 from Outdoor Sources." *Indoor Air* 24 (6), 639–651.

UK Government. 2008. *Climate Change Act 2008*. Chapter 27. Norwich: The Stationery Office.

WHO. 2010. *WHO Guidelines for Indoor Air Quality: Selected Pollutants*. Bonn: World Health Organisation European Centre for Environment and Health.

Wolfsense. 2014. *Utilizing PIDs (for VOCs) During IAQ Investigations. Application Note PID-IAQ-R2 8/14*. Shelton: Wolfsense.

Zero Carbon Hub. 2015. *Impact of Overheating. Evidence Review*. London: Zero Carbon Hub.

Estimating overheating in European dwellings

Luisa Brotas [ID] and Fergus Nicol [ID]

ABSTRACT

In recent years, the urgent need to adapt our lifestyles and buildings to deal with a more extreme and a warming climate has become clear, not least through the increasing overheating of buildings. This is reflected in the rising concerns about the discomfort and heat stress to building occupants caused by the increasing indoor temperatures. European standard BS15251 and Chartered Institution of Building Services Engineers (CIBSE) guidance note TM52 are documents that address the issue. Both include a methodology predicting the probability of overheating in buildings. Despite this, many modern buildings overheat. This paper looks at the criteria from CIBSE TM52 and discusses their applicability to a single UK dwelling archetype. This was modelled and then located in a range of European cities to understand the causes of overheating and the means of reducing it. Results highlight some problems in practice using simulations tools to evaluate overheating and the fundamental assumptions on which they are based. Energy performance and thermal comfort of dwellings were assessed using morphed climates for each location for 2020, 2050 and 2080.

Introduction

There is growing evidence that the global climate is warming and that this is a result of human activities. Increased levels of greenhouse gas emissions and other forms of environmental degradation such as the destruction of rainforests can speed global warming, and lead to an increased frequency of extreme weather events (IPCC 2014).

Concerted actions have been promoted to minimize man-made emissions to slow down the rate of global warming over the coming decades. The recent Climate Change summit COP21 held in Paris in December 2015 emphasized the urgent need to limit temperature rises well below 2 K – and if possible attempt to limit it to 1.5 K. This is an important climate deal that commits all countries to cut emissions when the Kyoto protocol comes to an end in 2020. Another factor in the deal is the requirement for nations to assess their progress towards meeting their climate commitments and submit proposals for the revision of their own defined targets (non-binding) every 5 years (COP21 2015). This may avoid the need to strike perennial deals and is expected to minimize delays in their implementation. The proposed schedule of regular revisions has been designed to facilitate the movement of governments and populations away from simple awareness to more concrete actions.

Green agenda

Buildings account for 40% of total energy consumption in the European Union (EPBD 2010). Therefore, it is imperative to reduce the use of fossil fuel energy that is contributing to greenhouse gas emissions, while promoting the use of renewables.

Recent energy supply uncertainty (threats of blackouts) and fuel price rises have been a driver for the promotion of nearly zero carbon buildings (ZCH 2009). Reducing heat losses, increasing energy efficiency and adopting renewable energy have been at the forefront of most EU regulations (EPBD 2010; EED 2012). However, an increasing use of new technologies and an interest in mechanical comfort cooling associated with global warming have counterbalanced the expected reduction of carbon emissions and in some cases have even aggravated their growth.

The Energy Performance of Buildings Directive (recast) has imposed that all new buildings should be nearly zero energy buildings from 2020 (EPBD 2010). The directive is to be transposed to the European member states who will individually define the parameters and the method they will use to achieve the European target of 20% reductions in energy and its resulting carbon emissions. The directive requires a 20% increase in energy efficiency and a 20% increase in the renewable sector, all to be achieved by the year 2020. In the UK, buildings are responsible for almost 50% of the country's carbon emissions. Ambitious plans for new dwellings to be zero carbon by 2016 were proposed by the UK government in alignment with the European Policy, but were later withdrawn.

However, European regulations still mainly address the heating season. Emphasis in them on reducing heat losses by building fabric and infiltration is promoting compact, lightweight, airtight buildings, increasingly reliant on mechanical ventilation albeit with heat recovery. Such solutions are effective at minimizing heating loads and providing comfortable temperatures in cold weather, but they can increase the risk of overheating impact in hot weather. Consideration of passive solutions for both heating and cooling needs to be taken in parallel, even in mild climates. Buildings and cities need to become resilient to more frequent extreme weather events and to heat waves, in particular.

Design solutions for new and refurbished buildings need to be low energy, while envisaging the present requirements and needs without compromising the needs in the future and as far as possible contributing to sustainable buildings and cities. At the same time, buildings must consider the comfort of the occupants and provide opportunities to restore or maintain acceptable environments. These are primary defences against the effects of climate change (Nicol, Humphreys, and Roaf 2012; Roaf, Brotas, and Nicol 2015). Furthermore, low-energy design becomes even more relevant in retaining stability in the energy supply. It is also important to take into account the needs of an ageing population and the problem of fuel poverty even in industrialized countries.

Regulations, Standards and Guidelines are good references to access and quantify the impact of changes in buildings. In Europe, EN15251 (BSI 2007) advises on indoor air quality, thermal environment, lighting and acoustics, suggesting the optimal values of various physical variables to use for energy calculations. EN15251 (and its impending redraft) looks at thermal comfort in free-running naturally ventilated buildings using the Adaptive Comfort approach. More recently, the Chartered Institution of Building Services Engineers (CIBSE) has produced Technical Memorandum (TM) 52 (2013) which provides the criteria by which to judge the risk of overheating in buildings. While more real data are still needed to validate these models, recent developments in dynamic building simulation software give an opportunity to test future scenarios.

Overheating in dwellings

Studies indicate that overheating is already a problem in a prototype tested across different climates in Europe (Brotas and Nicol 2015, 2016). This is also indicated by others (AECOM 2012; Lomas and Kane 2013; Mavrogianni et al. 2014, 2017; Psomas et al. 2016). There are also records which suggest that European dwellings which have been refurbished to improve the thermal performance in winter are now facing overheating problems in summer. This also occurs in new buildings designed to achieve the PassivHaus standard (Psomas et al. 2016). See also Lomas and Porritt (2017) for a thorough review of ongoing research on overheating in buildings.

See Figure 1 for a representation of the energy consumption for heating and cooling for a mid-floor flat in different countries in Europe and for climate predictions of 2020, 2050 and 2080 (Brotas and Nicol 2015). The flat which has been simulated is suggested by the Zero Carbon Hub as typical of such flats in the UK (ZCH 2009). While this model has high internal gains (assumed as representative of an increasing use of appliances in dwellings), this clearly highlights the predominance of cooling loads even in mild climates. This can be further exacerbated with climate change aggravated emissions scenarios and Urban Heat Island (UHI) phenomena (Kolokotroni et al. 2010; Lafuente and Brotas 2014; Santamouris 2014). The rapid urbanization of cities, the pressure associated with land costs and scarcity of space, has resulted in a more compact urban landscape with high and dense constructions and materials and less open/green spaces. All these factors can reduce or even prevent the possibility of adopting design solutions and strategies that can be effective in avoiding or reducing the need for mechanical systems for cooling. They can also reduce the proportion of the year during which either heating or cooling is required.

While dwellings have been less prone than other building types to adopting such active systems to deal with temperature rises, there is a growth in the sales of air-conditioning units or energy-inefficient cooling devices across Europe. Unprecedented recent heat waves particularly affecting vulnerable populations raise awareness of the impact of overheating on people's health (WHO 2009). Heat waves characterized by long duration and high intensity have the highest impact on mortality (WHO 2009; Santamouris and Kolokotsa 2015).

Method to assess the likelihood of overheating

The methodology suggested in European Norm and British Standard BS15251 (BSI 2007) to address the environmental parameters in naturally ventilated buildings is described in Nicol and Humphreys (2010). This was largely based in field studies undertaken in European cities, mostly in office buildings under the European project Smart Controls and Thermal Comfort (SCATs) (McCartney and Nicol 2002). This is acknowledged in the goals of the standard, but casts doubt on the suggestion that 'The standard is [thus] applicable to the following building types: single family houses, apartment buildings, offices, educational buildings', etc. (BSI 2007).

While its applicability in domestic buildings can be questioned, there is little evidence of contradictory indications. Oseland (1995) compared comfort perceptions in homes, offices and climate chambers and showed that people are less sensitive to temperature variations in their home. It also suggests that offering occupants control over temperature, through an interaction with the building (e.g. opening windows) or personal attitudes and behaviours (e.g. dress code), is the optimum strategy for energy efficiency and comfort in buildings. Nicol (2017) presents evidence that the limits for comfort in free-running residential buildings are typical of the general building stock, but suggests the use of mechanical temperature conditioning to control temperatures is used in a different way from that assumed for non-residential buildings.

The establishment of indoor environmental parameters for a chosen building system design and related energy performance calculations are typically developed according to a set of categories of buildings that are based on the expectations of the occupants in relation to their spatial requirements. This process, for instance, using TM52 differentiates free-running from mechanically ventilated buildings with each having different designated acceptable temperature ranges. Acceptable temperatures in free-running buildings are established using the 'adaptive method' already acknowledged in International standards (BSI 2007; ASHRAE, 2010). It suggests that the temperature that occupants will find comfortable or uncomfortable relates to the outdoor conditions in a predictable way. The range of acceptable indoor temperatures can be wider in free-running, naturally ventilated buildings where cooling is achieved by opening windows. The range of outdoor temperatures in which occupants

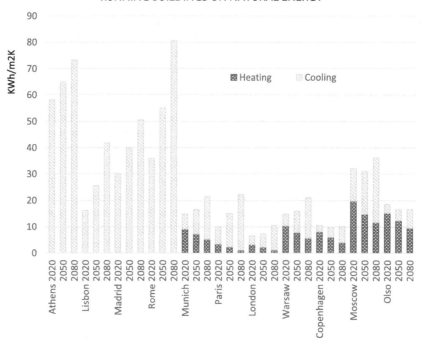

Figure 1. Heating and cooling energy consumption for dwellings in cities in Europe for 2020, 2050 and 2080.

can be comfortable indoors is narrower: as the use of mechanical heating and cooling allows occupants to be comfortable at higher or lower outdoor temperatures.

The TM52 (CIBSE 2013) uses both the adaptive and the steady state models (e.g. Predicted Mean Vote) to underpin its method when describing, defining and providing guidance on overheating to designers. A commonly used approach to identify overheating has been to look at the proportion of occupied hours with temperatures above a certain threshold.

In the UK, historic guidance from the superseded CIBSE Guide A (2006) was that overheating was likely to occur if operative temperatures exceeded 28°C for more than 1% of occupied hours. However, this limit was set irrespective of the outside conditions. Conditions for naturally ventilated buildings were set to be more flexible than in conditioned spaces, but occupant behaviour and the local climatic conditions were not fully considered. More recently, the CIBSE Guide (CIBSE 2015) suggests the use of the criteria proposed in TM52 (Nicol et al. 2009; CIBSE 2013).

The likelihood that a building will overheat can be predicted using monitoring or simulation. The simulation tool used should be able to use standard weather data sets to calculate the indoor Operative Temperature, T_{op}, and the Running Mean Outdoor Temperature, T_{rm}, if the adaptive model is to be used in the assessment. The model should include a realistic occupancy pattern of the building and the routine behaviour of the occupants within it. However, there is clearly a lot of uncertainty, especially in predictions developed using future climate scenarios. Other studies have highlighted problems associated with the reliability and variability of results for different climates, the morphing of future climates, the assessment of embodied energy and life cycle in the selection of solutions and materials, internal gains, occupancy profiles, operation modes, the performance gap between predicted and real consumption, and even the experience of the researcher in representing a model accurately

(de Wilde, Rafiq, and Beck 2008; Hacker et al. 2008; Taylor et al. 2014; Din and Brotas 2015, 2016; Symonds et al. 2017). These uncertainties are beyond the scope of this paper which presents a case study in various scenarios to discuss trends. It focuses on the pitfalls of trying to predict overheating in buildings and discusses solutions and criteria for a more thorough analysis in future studies.

TM52 overheating criteria

The criteria by which the danger of overheating can be assessed or identified in free-running buildings (those without mechanical heating or cooling) have been proposed based on previous studies, in particular research undertaken for the SCATs project in which a relationship between the indoor comfort temperature calculated from the data and the running mean of the outdoor temperature was derived from a broad survey of buildings in free-running mode (Nicol and Humphreys 2010):

$$T_c = 0.33T_{rm} + 18.8\,(°C), \tag{1}$$

where T_c is the predicted comfort temperature when the running mean of the outdoor temperature is T_{rm}.

According to TM52, the designer should aim at remaining within category II limits: Normal expectation (for new buildings and renovations) with a suggested acceptable temperature range of ± 3 K (BSI 2007)– updated in the EN16798 (2016) rewrite as $+3$ K for the upper limit of the indoor operative temperature and -4 K for the lower limit. Then from Equation (1), it follows that the upper limit of the range is T_{max}, where:

$$T_{max} = 0.33T_{rm} + 21.8\,(°C). \tag{2}$$

Simply exceeding T_{max} momentarily cannot be a reasonable justification to classify a building as overheating. Likewise, for criteria to be easily applied, overheating should be easily quantifiable.

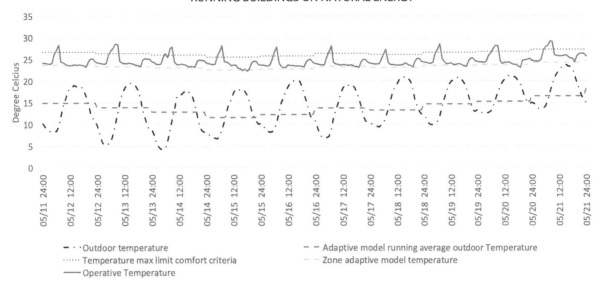

Figure 2. Sample of temperatures to assess overheating criteria in London 2030.

Figure 2 presents the operative temperature of a living room (with kitchen) and the comfort temperature and threshold upper limit comfort criteria. The operative temperature is clearly above the adaptive model temperature. But are the short daily periods where the indoor temperature is above the threshold enough to characterize this as an overheating space?

Notice that the amplitude of outdoor temperature is fairly regular. Yet, the adaptive running average outdoor temperature reaches its lowest of 12°C on 15/05 and increases by two or three degrees in a short period. The adaptive temperature is derived from this running mean and the temperature upper limit comfort criterion is the resulting temperature plus 3 K, originating a relatively constant threshold. Short peaks of the operative temperature will fail a limit of comfort.

The three criteria defined in TM52 (CIBSE 2013) propose a method by which the danger of overheating can be predicted. The method was developed by the CIBSE overheating task force and the description of the method given below is from CIBSE TM52 (CIBSE 2013).

The first criterion sets a limit for the number of occupied hours that the operative temperature can exceed T_{max} during a typical non-heating season, assumed for the TM52 as between 1 May and 30 September. The second criterion deals with the severity of overheating within any one day, which is given in terms of temperature rise and duration, and sets a daily limit for acceptability. The third criterion sets an absolute maximum acceptable temperature for a room (CIBSE 2013).

The criteria are all defined in terms of ΔT, the difference between the actual operative temperature in the room at any time, T_{op}, and T_{max}, the limiting maximum acceptable temperature calculated as shown in Equation (2). This ΔT is therefore:

$$\Delta T = T_{op} - T_{max}. \qquad (3)$$

ΔT is rounded to the nearest integer. ΔT will be negative unless the room is overheated. The three criteria for assessing whether a building is overheating are defined below.

Criterion 1: hours of exceedance (H_e)
The number of hours (H_e) during which ΔT is equal to or greater than one degree (K) shall be not more than 3% of occupied hours for the period of 1 May to 30 September.

This criterion can be used for a shorter period if data are not available or occupancy is only for a part of the period. The 3% should be based on the available period.

Criterion 2: daily weighted exceedance (W_e)
The severity of overheating is expressed as the weighted exceedance (W_e) that cannot exceed or be equal to six degree-hours in any day. This is expressed as:

$$W_e = \sum (h_e \times w_f)$$
$$= (h_{e0} \times 0) + (h_{e1} \times 1) + (h_{e2} \times 2) + (h_{e3} \times 3), \qquad (4)$$

where the weighting factor $w_f = 0$ if $\Delta T \leq 0$, otherwise, $w_f = \Delta T$, and h_{ey} is the number of hours when $w_f = y$.

An indication that the temperatures of a space are above a certain value may not be enough to assess the severity of overheating. If this exceedance is not significantly higher than the limit, it may not be perceived as discomfort, as adaptive mechanisms may take place. However, if this exceedance is significant or for a period longer than a few hours, it becomes problematic.

Criterion 3: upper limit temperature (T_{upp})
This criterion sets an absolute maximum indoor operative temperature (T_{upp}), where the value of ΔT ($T_{upp} - T_{max}$) must be less than or equal to 4 K.

$$(T_{upp} \leq T_{max} + 4). \qquad (5)$$

The threshold or upper limit temperature sets a limit above which normal adaptive actions will be insufficient to restore personal comfort. It is likely that most people will consider it 'too hot'. This criterion accounts for extreme hot weather conditions (CIBSE 2013).

Figures 3 and 4 show the maximum and upper thresholds for the operative temperature with reference to the running average outdoor air temperature. This provides a visual indication

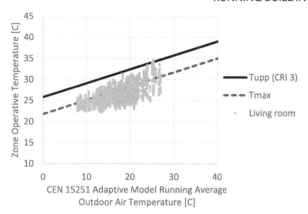

Figure 3. Likelihood of overheating with exceeding thresholds of Operative temperature versus Running Average Outdoor air temperature for Moscow in 2020.

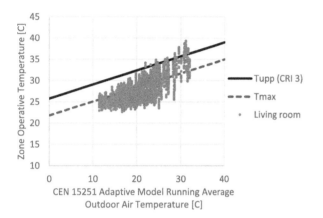

Figure 4. Likelihood of overheating with exceeding thresholds of Operative temperature versus Running Average Outdoor air temperature for Moscow in 2080.

of the range of temperatures achieved in a space as well as the likelihood of these exceeding the reference thresholds.

Quantifying the number of hours that the operative temperature indoors exceeds a maximum (T_{max}) and an upper limit (T_{upp}) for a certain period can provide an indication of whether the space is likely to be overheating. The 'hours of exceedance' criterion adopts a concept similar to the percentage of hours above a certain threshold previously suggested in CIBSE Guide A. The 3% maximum limit of occupied hours is suggested in BS EN15251.

The TM52 memorandum suggests its applicability across Europe. The data on which the adaptive standard developed were all collected from offices throughout Europe at all times of year. While it may be argued that the diversity of latitudes and climates across Europe may require a similar variety of periods of assessment, the five hottest months will give a reasonable indication of the likelihood of overheating in local buildings. The use of the 3% limit over the whole summer season has been queried by Lee and Steemers (2017), who suggest that overheating can be caused by much shorter periods.

A second aspect that seems relevant in the criteria is the method used to account for occupied hours. It is reasonable to expect that overheating in buildings is only a problem when it occurs in occupied periods. It is interesting to discuss the impact the occupancy pattern has on the assessment of the overheating criteria. Buildings that are occupied during daytime hours are more likely to fail the criteria as the assessment considers the

hotter period of the day. Likewise, assuming a permanent occupation of 24 hours all year will spread and may dilute the impact of a high percentage of overheating hours that occur during day. It is not unheard of that consultants may extend the occupancy profile for an hour or so to avoid a building being classified as overheating. Another situation that frequently occurs in domestic buildings is that they are unoccupied during daytime hours at least on week days. Adopting a predominantly night profile plus weekends can potentially reduce significantly the risk of failing criteria. Thinking about possible future trends towards working from home and of the ageing of populations, it is reasonable to assume a permanent occupation of domestic buildings with more than single occupancy. This was the adopted option in this study.

According to the above British Standard 15251 and TM52, the criteria to assess overheating are applicable when the running mean outdoor air temperature is between 10°C and 30°C during the period of assessment. The 30°C limit is related to the conditions applied in the data on which the standard is based. In mild climates, it is unlikely that the running mean outdoor temperature will exceed the upper threshold. Conversely, the lower range defines the need to heat, not cool. However, with global warming, temperatures may well begin to reach this limit in many European cities. Questions that may arise when assessing overheating include the following.

How to quantify the hours when the T_{rm} is above 30°C? Is this assumed as a threshold where active systems for cooling will need to be put in place? Then how is this moderated? Or shall it be assumed as an immediate overheating hour? This can easily be adopted for Criteria 1 and 3, but the question remains of how to quantify the range of exceedance for Criterion 2. Could one set an arbitrary limit of six degree-hours? Or should it be accounted in terms of ΔT, the difference between the indoor and the outdoor temperature? Results from the case study presented give some insight into such discussions.

Model

The present case study is located in the city of London. A base case model of a mid-storey flat (67 m²) is adopted based on statistics of housing stock broken down by type and in line with a rapid urbanization of cities.

The main façade is oriented east and the secondary faces north (see Figure 5). The thermal characteristics of the envelope comply with the Minimum Fabric Energy Efficient Standard (FEES) for 2016 from UK Part L1A (2013) regulations: external walls (U-value 0.18 W/m²K), party walls (0 W/m²K), semi-exposed wall (0.17 W/m²K) and windows (U-value 1.4 W/m²K and G-value 0.63). The layout and further details from the building envelope, ventilation and systems specifications are defined after Zero Carbon Homes (ZCH 2009, 2012). The occupancy profile is defined for three people in a domestic environment, assuming one person is permanently at home. This agrees with future trends towards home-working and an ageing population that may stay indoors most of the time. However, the assessment of overheating will assume in scenario 'day occup' a period of occupancy between 8 am and 6 pm for seven days in the week between 1 May and 30 September. This selected period is to account for the hotter period of the day. Scenario 'all' will assess

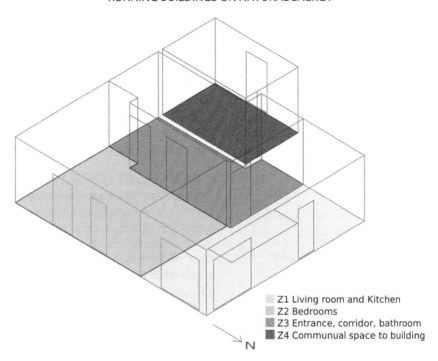

Z1 Living room and Kitchen
Z2 Bedrooms
Z3 Entrance, corridor, bathroom
Z4 Communual space to building

N

Figure 5. Wireframe model thermal zones. Z1 includes living room and kitchen; Z2 represents the bedrooms; Z3 includes entrance, corridor and bathrooms; and Z4 is external to the flat and models a communal space to the building (boundary condition assumed as an interior wall). Zones 1 and 2 have exterior walls and a party wall (modelled as adiabatic) in contact with a flat assumed with the same internal conditions.

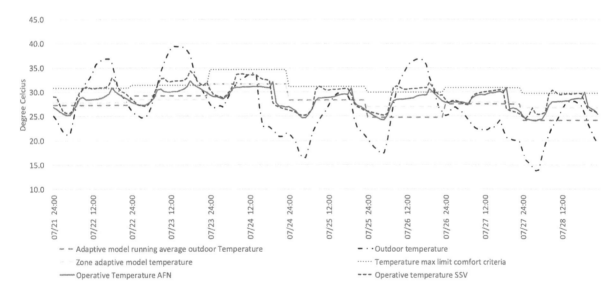

Figure 6. Extreme week for Munich for the year 2080. Operative temperature for model with cross ventilation (AFN) and single-side ventilation (SSV).

overheating on a 24-h occupancy for seven days a week. Note that this assessment is independent of the actual occupancy of the space. This is acknowledged as a conservative value for the periods the space is unoccupied, and will therefore attract lower internal gains.

The assessment of 24 hours distributes the impact of peak periods over longer hours. Conversely, it should be kept in mind that heat stored in high-density materials commonly found in cities can delay the cooling of the urban infrastructure of buildings for a couple of hours. This may mean that outdoor and connected indoor temperatures during early evening hours may be higher.

The initial base case assumes an infiltration specified as 0.3 ac/h as its design level. Single-sided night cooling ventilation (driven by wind and stack effect) that will influence individual rooms is activated when the indoor temperature is above 24°C and the delta differential to the outdoor is less than 2°C. An internal blind shading device with 0.1 visible and solar transmittance is activated when the indoor temperature rises above 24°C and solar radiation incident on the window is above 120 W/m². The thermal characteristics of the base model and adopted strategies are already energy efficient to a high standard.

The second combination of strategies was derived from previous related studies and is in line with the findings by other

Table 1. TM52 criteria for both models under analysis.

	C1	C2
	%	Days (6 degree-hours)
SSV	52	89
AFN	44	90

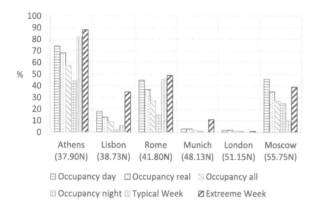

Figure 7. Criterion 1 (3% above threshold) for different locations and reference periods with the 2080 climate.

authors. Shading, ventilation and thermal mass are identified as first-line strategies to mitigate temperature rises in dwellings (Kolokotroni, Giannitsaris, and Watkins 2006; Gupta and Gregg 2012; AECOM 2012; Santamouris and Kolokotsa 2013; Mavrogianni et al. 2014). The second model assumes a cross ventilation defined as a multizone airflow network driven by wind direction, speed, orientation of the opening and temperature difference between indoors and outdoors. As before, the night cooling ventilation is activated when the indoor temperature is above 24°C and the delta differential to the outdoor is less than 2°C. While this replicates an automatic system, it can also represent to some extent occupants opening windows when they feel hot and leaving them open until the space cools down to feel comfortable, at temperatures defined in the model.

This model also integrates an external shading device with similar transmittance characteristics and operating profile as the previous internal device. The selection of solutions was based on realistic proposals that would not significantly interfere with land scarcity and high real state value in cities or could eventually be adopted in a refurbishment. No restrictions from listed areas were considered in the adoption of external devices. However, the window opening is reduced to 30% to account for security measures or opening just a fraction of the window.

A third model assumes internal gains being significantly reduced. Data for operation and power for lighting (low energy) and equipment (energy efficient) were retrieved from Guide A from CIBSE (2015). Internal gains are assumed for this model to be relatively low in line with the idea that appliances must become more efficient in the near future, if we are to achieve the CO_2 reduction targets for mitigating climate change (EED 2012).

All the dynamic simulations were made with EnergyPlus software, version 8.4.0. The criteria for the assessment of overheating developed in TM52 were compiled in a spreadsheet.

Figure 6 shows the operative temperature for model 1 assuming single-side ventilation (SSV) and an internal blind and for model 2 optimized with cross ventilation and an external shutter (AFN). In terms of the TM52 criteria, both models pass the criteria (see Table 1 for details).

Figure 7 presents the results for Criterion 1 for varying occupancy profiles, from 9 am till 6 pm, the real occupancy of that particular room (8–11 pm), a permanent 24-h schedule and a night occupancy from midnight till 8 am and 7 pm till midnight. A Typical Summer week (nearest average temperature for summer) and an Extreme Summer Week (nearest maximum temperature for summer) have been retrieved from the EnergyPlus weather files. They vary for the different climates presented (see Tables 2–7 for an indication of the dates). The results presented for these week periods are based on a 24-h permanent occupancy

Table 2. TM52 criteria for different periods for analysis in Athens.

Athens (37.90N)		C1	C2		C3
		%	Days (6 degree-hours)	Weeks (21 hours)	Hours
2020	Occupancy 9–18	52	89	14	119
	Day	44	90	16	138
	Occupancy all	34	90	16	138
	Night	20	45	9	19
	Typical week (29 June–5 July) all	51	6		8
	Extreme week (3–9 August) all	61	6		39
2050	Occupancy 9–18	63	105	18	311
	Day	55	105	19	378
	Occupancy all	44	105	19	391
	Night	31	70	13	80
	Typical week (29 June–5 July) all	67	7		29
	Extreme week (3–9 August) all	73	7		86
2080	Occupancy 9–18	74	121	20	536
	Day	68	122	20	703
	Occupancy all	57	122	20	802
	Night	44	88	16	266
	Typical week (29 June–5 July) all	82	7		59
	Extreme week (3–9 August) all	88	7		114

Note: Highlighted in dark grey are the TM52 criteria that pass for the day occupancy and in light grey are the misses.

Table 3. TM52 criteria for different periods for analysis in Lisbon.

Lisbon (38.73N)	C1 %	C2 Days (6 degree-hours)	C2 Weeks (21 hours)	C3 Hours
2020				
Occupancy 9–18	1	1	1	8
Day	1	2	1	10
Occupancy all	1	2	1	10
Night	0	1	0	2
Typical week (5–11 August) all	0	0		0
Extreme week (15–21 July) all	0	0		0
2050				
Occupancy 9–18	1	2	1	8
Day	1	2	1	10
Occupancy all	1	2	1	10
Night	0	1	0	2
Typical week (5–11 August) all	1	0		0
Extreme week (15–21 July) all	1	0		0
2080				
Occupancy 9–18	18	36	7	4
Day	13	37	7	37
Occupancy all	9	37	7	37
Night	2	3	0	1
Typical week (5–11 August) all	6	2		0
Extreme week (15–21 July) all	35	3		0

Note: Details as in Table 2.

Table 4. TM52 criteria for different periods for analysis in Rome.

Rome (41.80N)	C1 %	C2 Days (6 degree-hours)	C2 Weeks (21 hours)	C3 Hours
2020				
Occupancy 9–18	13	21	4	11
Day	10	23	4	11
Occupancy all	7	23	4	11
Night	4	6	2	0
Typical week (24–30 August) all	0	0		0
Extreme week (27 July–2 August) all	2	0		0
2050				
Occupancy 9–18	23	38	7	41
Day	18	41	7	45
Occupancy all	13	41	7	45
Night	7	12	2	4
Typical week (24–30 August) all	12	2		0
Extreme week (27 July–2 August) all	27	4		0
2080				
Occupancy 9–18	45	75	14	86
Day	37	78	16	104
Occupancy all	27	78	16	119
Night	15	27	8	33
Typical week (24–30 August) all	46	6		0
Extreme week (27 July–2 August) all	49	7		0

Note: Details as in Table 2.

(all). Longer hours of occupancy result in a reduction of the percentage of overheating. Moscow (Table 7) presents a reduced reduction between the permanent and the night occupancy. This may be a result of a lower daily amplitude or the impact of UHI.

Some weather files present very unusual peak temperature periods. These may compromise compliance with Criteria 2 and 3. The adoption of the adaptive running mean outdoor temperature may attenuate this phenomenon. There is also an ongoing discussion about whether Criterion 2 instead of a six hours/day should not be extended to a week period. This would avoid failing criteria on certain climates by a very small margin. This could eventually be twisted by selecting the start weekday of the simulation and planning for these peaks to fall on an unoccupied period (i.e. weekend for services). Using data

for a week rather than a day for the assessment of Criterion 2 would prevent these omissions and attenuate its impact for the criteria.

The typical and extreme weeks do not seem to present a consistent variation amongst the climates here presented, though there are some similarities of the week criterion assessment between Athens and Rome and between Lisbon, Munich and Moscow. The results from adopting a particular week in the summer period for Criterion 2 do not seem to be very consistent. Preliminary data not presented here further recognize this difficulty.

Figure 8 presents the daily weighted exceedance W_e for different occupancies at various locations in Europe with a climate file morphing 2080. Figure 9 presents a similar method but accounting for a week period of exceedance above 21 h.

Table 5. TM52 criteria for different periods for analysis in Munich.

Munich (48.13N)	C1	C2		C3
	%	Days (6 degree-hours)	Weeks (21 hours)	Hours
2020 Occupancy 9–18	2	2	0	0
Day	2	2	0	0
Occupancy all	1	2	0	0
Night	0	0	0	0
Typical week (15–21 July) all	0	0		0
Extreme week (22–28 July) all	1	0		0
2050 Occupancy 9–18	2	2	0	0
Day	2	2	0	0
Occupancy all	1	2	0	0
Night	0	0	0	0
Typical week (15–21 July) all	0	0		0
Extreme week (22–28 July) all	2	0		0
2080 Occupancy 9–18	3	4	0	0
Day	2	5	0	0
Occupancy all	2	5	0	0
Night	1	0	0	0
Typical week (15–21 July) all	0	0		0
Extreme week (22–28 July) all	11	0		0

Note: Details as in Table 2.

Table 6. TM52 criteria for different periods for analysis in London.

London (51.15N)	C1	C2		C3
	%	Days (6 degree-hours)	Weeks (21 hours)	Hours
2020 Occupancy 9–18	2	3	0	
Day	2	3	1	3
Occupancy all	1	3	1	3
Night	1	3	0	0
Typical week (29 June–5 July) all	0	0		0
Extreme week (17–23 August) All	0	0		0
2050 Occupancy 9–18	2	3	1	9
Day	2	3	1	9
Occupancy all	1	3	1	9
Night	1	2	1	0
Typical week (29 June–5 July) all	0	0		0
Extreme week (17–23 August) All	0	0		0
2080 Occupancy 9–18	2	3	1	10
Day	2	3	1	12
Occupancy all	1	3	1	12
Night	1	2	1	2
Typical week (29 June–5 July) all	0	0		0
Extreme week (17–23 August) All	1	0		0

Note: Details as in Table 2.

This approach is meant to simplify the collation of data and will account for possible periods that fall outside a typical or extreme reference week. A comparison between the two periods suggests that the daily period will be more sensitive to small variations. While a method should be relatively easy to apply (hence 24 h during 7 days), its accuracy is also important as it may compromise sensitivity analysis between different variables. This is important as the TM52 criteria can be a good mechanism to evaluate the impact of different design solutions to minimize overheating.

Figure 10 presents the results for Criterion 3 for different occupancy profiles. As before, caution should be taken to define the occupancy profile as it may strongly influence the validity of results obtained. Unlike the first two criteria, the third criterion tends to exaggerate the effect of extended hours of occupancy. This defines a maximum temperature limit threshold, beyond which adaptive opportunities may not be sufficient to restore comfort.

Tables 2–7 show the combination of the three TM52 criteria applied to a prototype simulated at various cities in Europe. While this particular model has been selected as representative of an overheating scenario in most of the climates, it also raises questions whether the three criteria would need to be checked. Likewise, can a building failing two criteria out of three for a small number of instances (Lisbon and London) be at risk of overheating? The representation of possible alternative approaches in terms of length of the period under assessment is the start of an ongoing project. It still needs a significant compilation of

Table 7. TM52 criteria for different periods for analysis in Moscow.

Moscow (55.75N)	C1 %	C2 Days (6 degree-hours)	C2 Weeks (21 hours)	C3 Hours
2020 Occupancy 9–18	26	45	8	3
Day	21	47	10	3
Occupancy all	15	47	10	3
Night	7	8	2	0
Typical week (6–12 July) all	11	2		0
Extreme week (29 June–5 July) all	26	3		0
2050 Occupancy 9–18	29	48	10	17
Day	25	52	13	22
Occupancy all	18	52	13	22
Night	10	14	4	5
Typical week (6–12 July) all	8	1		0
Extreme week (29 June–5 July) all	34	4		0
2080 Occupancy 9–18	46	62	12	93
Day	35	70	15	130
Occupancy all	27	70	15	143
Night	25	38	7	50
Typical week (6–12 July) all	10	1		0
Extreme week (29 June–5 July) all	39	5		1

Note: Details as in Table 2.

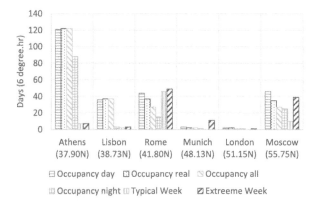

Figure 8. TM52 Criterion 2, number of days the daily temperature-weighted exceedance W_e is greater than 6 degree-hours.

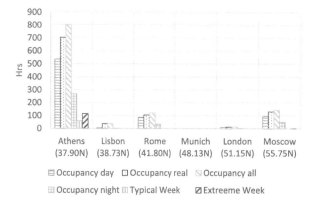

Figure 10. Criterion 3 for different occupancy solutions at different city locations across Europe for the climate of 2080.

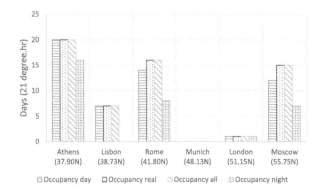

Figure 9. Overheating criterion over a week period when it exceeds 21 degree-hours, for different locations across Europe for the 2080 climate.

different simulation models or monitoring data to derive sensible conclusions.

Conclusions

Climate change is creating more and more potentially devastating and unpredictable warming events. A clear green agenda is required to build resilience to the impacts of such global trends that takes into account the need to both mitigate against climate change by reducing emissions, and adapt to it by instigating design practices that minimize the likelihood of overheating during hot weather events (COP21 2015). Consequently, there is a strong imperative to put the need to prevent overheating in building high on designers' agendas to both reduce thermal dissatisfaction with the built environment and obviate the drift towards the use of mechanical cooling to counteract growing levels of overheating in buildings.

Overheating is already a growing problem in locations and building types across Europe. It is therefore relevant to clearly identify a common methodology to assess its magnitude. The criteria developed by the overheating task force set up by CIBSE and introduced in TM52 are based on the adaptive comfort method. The relationship of the indoor operative temperature to the comfort temperature calculated from the running outdoor mean temperature as suggested by European Standard EN15251 seems to be a step forward when compared to previous methods based on a fixed temperature threshold.

This paper raises some problems associated with the application of the TM52 criteria in practice using simulation tools to evaluate overheating and the fundamental assumptions on which they are based. Some of the problems encountered using the TM52 methodology in this study include the following:

(1) The different time periods used in the assessment of the three criteria – In Criterion 1, the assessment is made over the whole of the cooling season, while in Criteria 2 and 3, a single day of non-compliance can be seen as enough to rule the building to be overheating. This study indicates that a time period of a week might be more realistic, at least for Criterion 2. This requires a change in the acceptable degree-hour exceedance from six degree -hours in any one day as suggested in TM52 to 21 degree-hours in any one week. This kind of change to Criterion 2 has been suggested elsewhere (e.g. suggested rewriting of comfort criteria in BS101).

(2) A clear flaw in the method relates to increasing global temperatures and the arbitrary limit of applicability of the criteria set at 30°C. This limit was originally set simply because of the limited range of temperatures that occurred during the SCATs project that resulted in the database on which the method is based (McCartney and Nicol 2002). This database is used to calculate the relationship between comfort temperatures and outdoor temperature in TM52 (Nicol and Humphreys 2010). Similar approaches have been developed in the USA by de Dear and Brager (2001) and in hot climates such as that of Pakistan (Nicol et al. 1999) which suggest a broader range of temperatures that are currently applicable and will be even more so in the warmer future. Except for extreme heat wave events, only some of the climates in Europe currently reach such high temperatures, but the rate at which this may change is unpredictable. Nevertheless, climatic file predictions up to the year 2080 are already affected by this limitation in the method, which is being increasingly used when assessing the long-term impacts of solutions. It is timely to address these questions now with some urgency. Adaptive opportunities and climatic adaptation may mean that in the future Europeans may well behave as other populations from hotter climates already do. This will then mean the formula suggested in BS15251 and TM52 will need to be extended, and in light of the record hot years experienced since the turn of the century, the sooner this is done the better.

(3) The development of the adaptive comfort model in EN15251 (and consequently in TM52 which derives from it) includes an inbuilt assumption that comfort and discomfort arise overwhelmingly from the thermal environment. As a result, the reality of discomfort as a social and psychological construct is not fully acknowledged, and hence the meaning of the term 'overheating' has never been fully explored. To some extent, this paper solves this problem by using the basic behavioural adaptive approach to thermal comfort. Hopefully, future research will develop a more rounded nuanced picture of overheating in buildings.

Disclosure statement

No potential conflict of interest was reported by the authors.

ORCID

Luisa Brotas http://orcid.org/0000-0001-7521-4474
Fergus Nicol http://orcid.org/0000-0003-0035-7918

References

AECOM. 2012. *Investigation into Overheating in Homes. Analysis of Gaps and Recommendations*. London: Publications Department for Communities and Local Government.

ASHRAE. 2010. *ANSI/ASHRAE Standard 55-2010: Thermal Environmental Conditions for Human Occupancy Atlanta*. Georgia: American Society of Heating, Refrigerating and Air-Conditioning Engineers.

Brotas, L., and F. Nicol. 2015. "Adaptive Comfort Model and Overheating in Europe Re-Thinking the Future." 31th International PLEA Conference Architecture in (R)evolution Proceedings. Ass. Building Green Futures, Bologna, Italy, 31th International PLEA Conference Architecture in (R)evolution, Bologna, Italy, September 9–11.

Brotas, L., and F. Nicol. 2016. "Using Passive Strategies to Prevent Overheating and Promote Resilient Buildings." In *PLEA 2016 – Cities, Buildings, People: Towards Regenerative Environments, In Proceedings of the 32nd International Conference on Passive and Low Energy Architecture*, edited by Pablo LaRoche and Marc Schiler, Los Angeles, USA, 135–143. ISBN 978-0-692-74961-6, Session B4.

BSI. 2007. *BS EN15251: 2007 Indoor Environmental Input Parameters for Design and Assessment of Energy Performance of Buildings Addressing Indoor air Quality, Thermal Environment, Lighting and Acoustics*. London: BSI.

CIBSE. 2015. *Environmental Criteria for Design. Chapter 1 in CIBSE Guide A Environmental Design*. London: Chartered Institution of Building Services Engineers.

CIBSE TM52. 2013. "The Limits of Thermal Comfort: Avoiding Overheating in European Buildings." CIBSE Technical Memorandum No 52. London: The Chartered Institution of Building Services Engineers. July 2013.

COP21. 2015. "Adoption of the Paris Agreement. Proposal by the President Draft decision CP.21." Report FCCC/CP/2015/L.9/Rev.1 [Online], United Nations and Framework convention on Climate Change. Paris. Accessed 8 January 2016. https://unfccc.int/resource/docs/2015/cop21/eng/l09r01.pdf.

de Dear, R. J., and G. S. Brager. 2001. "The Adaptive Model of Thermal Comfort and Energy Conservation in the Built Environment." *International Journal of Biometeorology* 45 (2): 100–108.

Din, A., and L. Brotas. 2015. "The LCA Impact of Thermal Mass on Overheating in UK Under Future Climates." PLEA 2015 – 31th International PLEA Conference: ARCHITECTURE IN (R)EVOLUTION, Bologna, September 9–11, ID 443.

Din, A., and L. Brotas. 2016. "Exploration of Life Cycle Data Calculation: Lessons From a Passivhaus Case Study." *Energy and Buildings* 118: 82–92. doi:10.1016/j.enbuild.2016.02.032.

EED. 2012. Energy efficiency. Directive 2012/27/EU of the European Parliament and of the Council, October 25.

EPBD. 2010. Energy performance of buildings (recast). Directive 2010/31/EU of the European Parliament and of the Council, May 19.

Gupta, R., and M. Gregg. 2012. "Using UK Climate Change Projections to Adapt Existing English Homes for a Warming Climate." *Building and Environment* 55: 20–42.

Hacker, J. N., T. P. De Saulles, A. J. Minson, and M. J. Holmes. 2008. "Embodied and Operational Carbon Dioxide Emissions From Housing: A Case Study on the Effects of Thermal Mass and Climate Change." *Energy and Buildings* 40: 375–384. doi:10.1016/j.enbuild.2007.03.005.

IPCC Fifth Assessment Synthesis Report. 2014. *Climate Change 2014 Synthesis Report.* [Online]. Accessed 5 February 2015. www.ipcc.ch/report/ar5/.

Kolokotroni, M., M. Davies, B. Croxford, S. Bhuiyan, and A. Mavrogianni. 2010. "A Validated Methodology for the Prediction of Heating and Cooling Energy Demand for Buildings Within the Urban Heat Island: Case-Study of London." *Solar Energy* 84: 2246–2255.

Kolokotroni, M., I. Giannitsaris, and R. Watkins. 2006. "The Effect of the London Urban Heat Island on Building Summer Cooling Demand and Night Ventilation Strategies." *Solar Energy* 80: 383–392.

Lafuente, J., and L. Brotas. 2014. "The Impact of the Urban Heat Island in the Energy Consumption and Overheating rof Domestic Buildings in London." Proceedings of 8th Windsor Conference: Counting the Cost of Comfort in a Changing World Cumberland Lodge, Windsor, UK, April 10–13, 805–815. London: Network for Comfort and Energy Use in Buildings. http://nceub.org.uk, ISBN 978-0-9928957-0-9.

Lee, V., and K. Steemers. 2017. "Exposure Duration in Overheating Assessments: a Retrofit Modelling Study." *Building Research & Information* 45 (1–2): 60–82. doi:10.1080/09613218.2017.1252614.

Lomas, K. J., and T. Kane. 2013. "Summertime Temperatures and Thermal Comfort in UK Homes." *Building Research & Information* 41 (3): 259–280.

Lomas, K. J., and S. M. Porritt. 2017. "Overheating in Buildings: Lessons From Research." *Building Research & Information* 45 (1–2): 1–18. doi:10.1080/09613218.2017.1256136.

Mavrogianni, A., M. Davies, J. Taylor, Z. Chalabi, P. Biddulph, E. Oikonomou, P. Das, and B. Jones. 2014. "The Impact of Occupancy Patterns, Occupant-Controlled Ventilation and Shading on Indoor Overheating Risk in Domestic Environments." *Building and Environment* 78: 183–198.

Mavrogianni, A., A. Pathan, E. Oikonomou, P. Biddulph, P. Symonds, and M. Davies. 2017. "Inhabitant Actions and Summer Overheating Risk in London Dwellings." *Building Research & Information,* 45 (1–2): 119–114. doi:10.1080/09613218.2016.1208431.

McCartney, K. J., and J. F. Nicol. 2002. "Developing an Adaptive Control Algorithm for Europe: Results of the SCATs Project." *Energy and Buildings.* 34 (6): 623–635.

Nicol, J. F. 2017. "Temperature and Adaptive Comfort in Heated, Cooled and Free-Running Dwellings." *Building Research & Information.* doi:10.1080/961328.2017.301698.

Nicol, F., J. Hacker, B. Spires, and H. Davies. 2009. "Suggestion for New Approach to Overheating Diagnostics." *Building Research & Information* 37: 348–357.

Nicol, J. F., and M. A. Humphreys. 2010. "Derivation of the Adaptive Equations for Thermal Comfort in Free-Running Buildings in European Standard EN15251." *Buildings and Environment* 45 (1): 11–17.

Nicol, F., M. Humphreys, and S. Roaf. 2012. *Adaptive Thermal Comfort: Principles and Practice.* Oxon and NY: Earthscan from Routhlege.

Nicol, J. F., I. A. Raja, A. Allaudin, and G. N. Jamy. 1999. "Climatic Variations in Comfortable Temperatures: the Pakistan Projects." *Energy and Buildings* 30 (3): 261–279. ISSN 0378-7788.

Oseland, N. A. 1995. "Predicted and Reported Thermal Sensation in Climate Chambers, Offices and Homes." *Energy and Buildings* 23 (2): 105–115.

Part L1A. 2013. *Approved Document L1A: Conservation of Fuel and Power in New Dwellings.* 2013 edition. London: NBS, part of RIBA Enterprises. ISBN 978-1-85946-510-3.

Psomas, T., P. Heiselberg, K. Duer, and E. Bjørn. 2016. "Overheating Risk Barriers to Energy Renovations of Single Family Houses: Multicriteria Analysis and Assessment." *Energy and Buildings* 117: 138–148.

Roaf, S., L. Brotas, and F. Nicol. 2015. "Counting the Costs of Comfort." *Building Research & Information* 43 (3): 269–273. doi:10.1080/09613218.2014. 998948.

Santamouris, M. 2014. "On the Energy Impact of Urban Heat Island and Global Warming on Buildings." *Energy and Buildings* 82: 100–113.

Santamouris, M., and D. Kolokotsa. 2013. "Passive Cooling Dissipation Techniques for Buildings and Other Structures: The State of the art." *Energy and Buildings* 57: 74–94.

Santamouris, M., and D. Kolokotsa. 2015. "On the Impact of Urban Overheating and Extreme Climatic Conditions on Housing, Energy, Comfort and Environmental Quality of Vulnerable Population in Europe." *Energy and Buildings* 98: 125–133.

Symonds, P., J. Taylor, A. Mavrogianni, M. Davies, c. Shrubsole, I. Hamilton, and Z. Chalabi. 2017. "Overheating in English Dwellings: Comparing Modelled and Monitored Large-Scale Datasets." *Building Research & Information* 45 (1–2): 195–208. doi:10.1080/09613218. 2016.1224675.

Taylor, J., M. Davies, A. Mavrogianni, Z. Chalabi, P. Biddulph, E. Oikonomou, P. Das, and B. Jones. 2014. "The Relative Importance of Input Weather Data for Indoor Overheating Risk Assessment in Dwellings." *Building and Environment,* 76: 81–91.

WHO. 2009. "Improving Public Health Responses to Extreme Weather/Heat-Waves – EuroHEAT." Technical summary. World Health Organization.

de Wilde, P., Y. Rafiq, and M. Beck. 2008. "Uncertainties in Predicting the Impact of Climate Change on Thermal Performance of Domestic Buildings in the UK." *Building Services Engineering Research and Technology* 29 (1): 7–26.

ZCH. 2009. *Defining a Fabric Energy Efficiency Standard for Zero Carbon Homes.* Milton Keynes: Zero Carbon Hub, [Online]. Accessed 2 May 2015. http://www.zerocarbonhub.org/resources.aspx.

ZCH. 2012. *Overheating in Homes: An Introduction for Planners, Designers and Property Owners.* Milton Keynes: Zero Carbon Hub.

Experimental validation of simulation and measurement-based overheating assessment approaches for residential buildings

Raimo Simson ⓘD, Jarek Kurnitski ⓘD and Kalle Kuusk

ABSTRACT

As a part of the building design process, Estonian building code requires standardized dynamic hourly simulations to verify the building's compliance to the summer thermal comfort requirements. In this study, we analysed this overheating assessment method for free-running residential buildings, by comparing the simulation results with measured data. Simulation models with different thermal zoning levels were studied: single-zone models, multi-zone apartment models and multi-zone whole building models. We analysed and quantified the effects of modelling detail and thermal zoning on indoor temperature and overheating estimation on the basis of five apartment buildings. Based on the results, a method, using indoor temperature measurements and outdoor climate data, to assess overheating risk has been proposed, as a relatively simple and inexpensive method for pre-defining the need for dynamic simulations.

1. Introduction

During the past decades, computer-based building modelling and simulation has become a common practice among engineers and architects (Attia et al. 2012), mainly with a goal to estimate the energy consumption of planned buildings. With the rapid evolution of building simulation tools and increase in computing power, more detailed and advanced models can be created and analysed, to imitate real buildings in operation. Aside from energy consumption estimation, accurate and detailed simulations of indoor climate parameters have been made possible (Wang and Zhai 2016).

With the trends in architecture and envelope design, namely the extensive usage of unshaded glazed façades and highly insulated airtight walls, an increasing number of low-energy buildings are built with a tendency to overheat (Mavrogianni et al. 2012; Chvatal and Corvacho 2009). These cases occur not only in warm climate countries but also in temperate and cold climate countries (Rohdin, Molin, and Moshfegh 2014).

Assessing the risk of overheating in buildings is rather a difficult and time-consuming task. The use of detailed dynamic simulations is becoming the mainstream method practised among architects and specialists, with also raising trends in analysing free-running domestic buildings. There are, however, numerous important variables causing differences between real situation and assessment results, such as occupant behaviour (Haldi and Robinson 2011), occupancy density and patterns in terms of internal gains, opening and closing windows (Schweiker et al. 2012), shading and air movement dynamics, which are difficult to predict (da Silva, Leal, and Andersen 2015). To reduce the complexity of such analysis, some forms of standardized methods are practised in different parts of the world (Jenkins et al. 2013). For example, in the UK, a simplified static calculation

assessment method can be used for residential developments (Tillson, Oreszczyn, and Palmer 2013; DECC 2014; Jenkins et al. 2013), in Finland on the other hand, multi-zone dynamic simulations are required by the Building Code (2012). Using the more complicated simulations to predict overheating with acceptable accuracy requires sufficiently detailed modelling with adequately defined thermal zoning, especially in case of low-energy and free-running buildings (O'Brien, Athienitis, and Kesik 2011). Simplifications in such thermal modelling and calculations are, of course, welcomed among building professionals (Kanters, Dubois, and Wall 2013), but can only to be stretched to a reasonable extent, in order to estimate building performance with an acceptable margin of error. Drawbacks of using such simplified approaches, in practice, have been also recently reported (Bateson 2016; Jenkins et al. 2013).

With the launch of European Union Directive 2010/31/EU (EU 2010), requirements for overheating prevention were established also for all new buildings in Estonia. According to the enforced Regulation No. 68 (GOV 2012b), the compliance verification calculation for summer thermal comfort in residential buildings needs to be conducted for at least one living room and one bedroom, with the highest risk of overheating. As opposed to the Finnish multi-zone methodology, for example, the Estonian approach implies single room calculations, in which the heat and air transfer dynamics of the apartment or the building as a whole are not accounted.

In the recent years, several studies have been carried out in Estonia on indoor climate in apartment buildings, mostly involving buildings built before 1990s, but also on newer buildings constructed between 1990 and 2010 (Kalamees et al. 2012). It was found that 63% of the studied dwellings were overheating in summer. Maivel, Kurnitski, and Kalamees (2014) investigated indoor temperature-related problems in old and new apartment buildings in Estonia and found that overheating is most common in new buildings. It was also concluded that for a detailed analysis, dynamic simulations are needed.

Simson, Kurnitski, and Maivel (2016) studied the current situation regarding compliance to summer thermal comfort, and found that 68% of the new buildings built in Estonia, after the enforcement of the new regulation, did not comply with the requirements.

The aim of the study is to analyse the impact of thermal zoning on the simulation-based overheating assessment calculation and to give a temperature measurement-based 'rule of thumb' for a low-cost method for pre-assessing overheating compliance of dwellings. We have compared measured hourly average indoor temperature with results from three levels of thermal zoning – the currently used single-zone method and two multizone approaches: whole apartment and whole building model approach. For detailed analysis, we have selected apartments from five apartment buildings in which temperature measurements have been conducted during the summer period from 1 July to 31 August 2014. For simulations, we used the energy and indoor climate simulation software IDA-ICE (EQUA 2016). In order to compare the calculation methods for summer thermal comfort assessment, we have fitted the simulation results using the temperature measurements.

2. Methods

First, whole building models were created and simulated with well-validated software IDA-ICE (EQUA 2016), using weather data from the year of the measurements. The simulated temperature results were compared with measured indoor temperature results. Then, the models were adjusted to get acceptable margin of error and correlation with measurements. These 'fitted' models were then simulated using weather data from Test Reference Year (TRY; Kalamees and Kurnitski 2006), to get a base value for temperature excess and evaluate the buildings' compliance with overheating requirements. To analyse the impact of thermal zoning, two alternative simulation models were created by removing neighbouring zones from the original whole building model: for the first model, only the rooms in the apartment were kept and for the second model, only the analysed rooms were kept. Simulation results from the latter models were also compared with the results from the fitted whole building model.

Based on the respective simulation results using real weather data, base temperature values for temperature excess calculations were calculated for each building to get respective excess values with measurement results.

2.1. Compliance assessment

The mandatory summer thermal comfort compliance verification in Estonia, for planned buildings, is carried out according to the requirements described in Estonian Regulations No. 63, 'Minimum Requirements of Energy Performance' (GOV 2012b) and No. 68, 'Methodology for Calculating the Energy Performance of Buildings' (GOV 2012a) using dynamic computer simulations. The methodology states that overheating risk assessment has to be done for living rooms and bedrooms, with the highest

potential to counter high temperatures. To quantify the overheating risk in these rooms, indoor temperature excess (DH) in degree-hours (Kh) is used, which is calculated as follows:

$$DH_{t_b} = \sum_{i=1}^{j} (t_i - t_b)^+, \tag{1}$$

where DH_{t_b} is the temperature excess in degree-hours over the base temperature t_b, t_i is the hourly mean room temperature and j is the total number of hours in the given period. The '+' sign means that only positive values are summed. In the Estonian regulation, the maximum limit for residential buildings indoor temperature excess is 150 Kh over a base temperature $t_b = +27°C$. For the calculation period of $j = 2208$ h, that is, from 1 July to 31 August, the equation can be given as follows:

$$DH_{+27°C} = \sum_{i=1}^{2208} (t_i - 27)^+. \tag{2}$$

2.2. Weather data

According to the methodology, room temperature calculations are performed regardless of the location of the building using the Estonian TRY (Kalamees and Kurnitski 2006), also used for building energy consumption calculations. The TRY is constructed using different months from three decades (1970–2000) of climatic data that best describe Estonian climate. It contains hourly mean data of outdoor temperature, relative humidity, wind speeds and solar radiation.

The indoor temperature measurements in dwellings were performed in the summer of 2014. Compared to outdoor temperatures from TRY and a typical summer of 2013 (Figure 1), the summer of 2014 was relatively warm, with two distinctive heat waves with hourly mean outdoor temperatures reaching higher than +30°C. The outdoor DH over the base temperature +27°C in 2013 was 24.3 Kh, in 2014 157.3 Kh. In the case of TRY, the DH is 0.5 Kh (Figure 1). For the measurement year 2014, a custom climate file was created using the measured data from a nearby weather station.

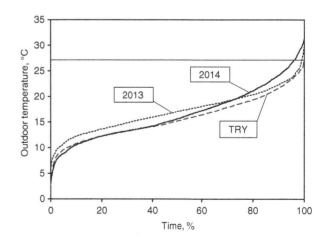

Figure 1. Hourly mean outdoor temperature duration curves for summer period from 1 July to 31 August. Data from Estonian TRY and weather station measurements [Estonian Weather Service (EMHI 2015)] in years 2013 and 2014.

2.3. Description of the studied buildings

Five apartments from five different buildings were studied, modelled and simulated. The relevant information for building structures, dimensions, site and other parameters was acquired from buildings' design documentation. Overview of the specifications of external boundaries, windows and other parameters of the buildings is given in Table 1. The studied buildings were constructed between 2011 and 2014. From each building, one apartment was selected for the analysis. Example plans and analysed rooms of the apartments are shown in Figure 2. All the buildings had apartment-based mechanical supply and exhaust ventilation units installed. Outdoor air was supplied to living rooms and bedrooms and removed from bathrooms and kitchens. The air handling units were equipped with summer bypass function for the heat exchanger. During the summer period, no heating systems were utilized in the buildings. Also, no mechanical cooling systems were installed (Figure 3).

2.4. Measurements

Indoor temperature measurements were carried out during the summer period of 1 July to 31 August 2014 in either living rooms or bedrooms in the selected apartments. For measurements previously calibrated, data logging Hobo U12 (ONSET 2015) devices were used with temperature measuring range of -20 to $+70°C$ with accuracy $\pm 0.35\,K$ and relative humidity 5–95% with accuracy $\pm 2.5\%$ of full-scale output. The data loggers were placed in the occupied zone of the rooms, away from direct sunlight, ventilation air flows, heat-generating equipment, etc. For each measurement taken, correction factors according to calibration results were applied. Ventilation air flows from supply and exhaust valves and grilles were measured with SwemaFlow 234 air flow hood with a range of 2–65 l/s and uncertainty $\pm 2.5\%$ of read value.

2.5. Simulations

The buildings were modelled with indoor climate and the energy simulation software IDA Indoor Climate and Energy, version 4.7 (IDA-ICE) (EQUA 2016). This tool allows detailed and dynamic whole-year multi-zone building simulations of indoor climate, energy consumption and building systems performance. IDA-ICE has been validated according to European Standard EN-ISO 13791 defined test cases (Kropf and Zweifel 2002), to Envelope BESTEST in the scope of IEA Task 12 (Achermann 2000) and used in several similar studies (Hamdy, Hasan, and Siren 2011; Hilliaho, Landensivu, and Vinha 2015; Jokisalo and Kurnitski 2007; Molin, Rohdin, and Moshfegh 2011).

Input data for the studied buildings, including building site surroundings, architecture, floor plans, and specifications of walls, roofs and windows were acquired from design documentation of the buildings and Estonian Registry of Buildings database (ERBD 2014).

Each material layer included properties for specific heat and density for accurate calculation of building thermal mass. Solar heat gain coefficients of windows, if not available in design documentation, were calculated using detailed window model with glazing properties calculation tool in IDA-ICE. Overall values used in buildings' simulations are shown in Table 1.

Trees, casting shades on the building, were modelled as crossing vertical rectangular planes (Table 1). The shading effect of trees and foliage was estimated as transparency factor between 0.2 and 0.3 (with 1.0 being fully transparent) (Heisler 1986).

Infiltration for the buildings was calculated using Equation (3) (GOV 2012a):

$$q_i = q_{50} \times A/(3.6 \times z), \tag{3}$$

where q_{50} is the building air permeability at 50 Pa pressure difference, $m^3/(h\cdot m^2)$; A is the total area of building envelope, m^2;

Table 1. Specifications overview of the studied buildings.

Building no.	B1	B2	B3	B4	B5
Photo of the studied building					
3D view of the building model in IDA-ICE					
Construction year	2014	2012	2011	2012	2013
Envelope construction	Concrete	Concrete	Pre-cast concrete	Pre-cast concrete	Concrete block
Building height(m)	14.0	11.7	21.0	12.0	10.6
Floors above ground	4	4	6	3	3
Apartments	12	21	40	14	9
Net heated area(m^2)	1 137	1 580	3 114	891	742
Volume(m^3)	5 465	6 043	11 422	4 872	2 884
Ext. wall U-value(W/($m^2\cdot K$))	0.20	0.16	0.17	0.21	0.19
Roof U-value(W/($m^2\cdot K$))	0.12	0.14	0.09	0.12	0.13
Windows U-value(W/($m^2\cdot K$))	1.10	1.00	0.89	1.10	1.20
Windows g-value	0.65	0.45	0.60	0.63	0.67

(a)

(b)

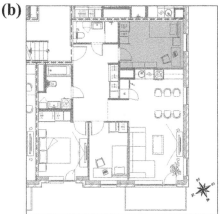

(c)

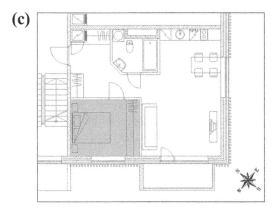

Figure 2. Example of apartment plans and analysed rooms (highlighted) of the studied buildings: bedroom, B1 (a); bedroom, B2 (b) and bedroom, B4 (c).

and z is the building height factor: 35 for one, 24 for two, 20 for three and four and 15 for five and higher storey buildings. In all cases, building air permeability value 3 m³/(h·m²) was used, as intended for calculations in new buildings, according to GOV (2012a).

The opening and closing of windows was modelled using on/off temperature control macro with a dead band of 2 K (Figure 4). This means that windows would open, when room temperature raised 1 K above the set point temperature value, and close, when dropped 1 K under the set point value. In this case, the set point was chosen +22°C, ensuring window openings at +23°C, and closings at room temperatures under +21°C. When outdoor temperature would exceed indoor temperature values, the windows would also close. The openable area was calculated as a percentage of the openable window total area, depending on the height and width of the window, imitating the airing position.

The regulation (GOV 2012a) gives values for the whole dwelling's internal gains as follows: 28.3 m² floor area per occupant with heat emission of 125 W, including 85 W sensible heat; 3 W/m² accounting equipment and 8 W/m² for lighting. The occupancy and load profiles are shown in Figure 5. Ventilation in zones was modelled as well mixed with constant supply and exhaust air flow rate of 0.5 l/(s·m²) (GOV 2012b). In apartments with central mechanical exhaust ventilation, the supply air temperature was taken equal to outdoor temperature. In apartments with local air handling units (AHUs), supply air

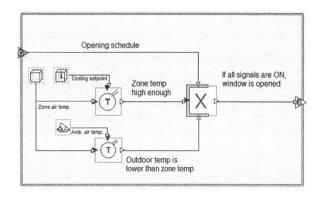

Figure 4. Window opening control macro in IDA-ICE used in simulations. Window opens if zone temperature exceeds cooling setpoint $t_{cool} + \Delta t/2$, and outdoor temperature is lower than room temperature, window closes when the zone temperature drops below $t_{cool} - \Delta t/2$. Δt is defined as dead band value.

Figure 3. Photos from the studied rooms: (from left) bedroom, building B1; bedroom, building B2; living room, building B3; living room, building B4 and bedroom, building B5.

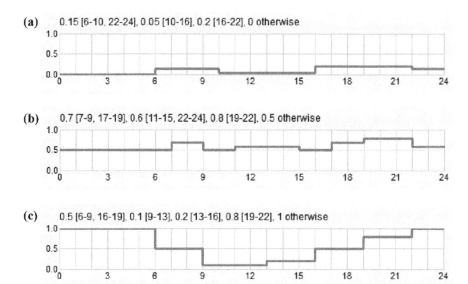

(a) 0.15 [6-10, 22-24], 0.05 [10-16], 0.2 [16-22], 0 otherwise

(b) 0.7 [7-9, 17-19], 0.6 [11-15, 22-24], 0.8 [19-22], 0.5 otherwise

(c) 0.5 [6-9, 16-19], 0.1 [9-13], 0.2 [13-16], 0.8 [19-22], 1 otherwise

Figure 5. Hourly profiles for lighting (a), equipment (b) and occupancy (c) used for internal heat gains calculation according to Estonian Regulation No. 63 (GOV 2012a).

temperature was considered 1 K higher than outdoor temperature due to the supply fan heat emission. Considering the bypass option of domestic AHUs, heat exchanger effect was not accounted.

2.5.1. Thermal zoning

We used three different thermal zoning approaches for building modelling (Figure 6):

- multi-zone approach, with all the rooms in the building modelled;
- multi-zone approach, with only rooms in the apartment modelled;
- single -zone approach, with only the analysed room modelled.

In case of the apartment-based method, thermal connections, as well as air leakages between other rooms and neighbouring apartments, openings and boundaries were not accounted – heat and air transfer was modelled only between the rooms in the apartment and outdoor environment, for example, external walls, internal walls and windows. The single-zone method accounted only for connections with external walls and windows, and the neighbouring sides of internal constructions were modelled as adiabatic. The multi-zone method, however, accounted connections between all the rooms in the apartment. In the single-zone method, both supply and exhaust ventilation was modelled as room-based.

An example schematic view of the whole building model in IDA-ICE simulation environment (SE) is presented in Figure 7. The IDA-ICE SE is a general-purpose modelling and simulation tool for modular systems where components are described with mathematical equations, written in the Neutral Model Format. More detailed information, for the component modules and IDA solver, can be obtained from several publications (Vuolle and Sahlin 2000; Björsell et al. 1999; Vuolle and Sahlin 1999; Kalamees 2004; EQUA 2016). The largest whole building model, with 153 thermal zones, was created for building B3, which had 40 apartments.

2.6. Evaluation of simulation results

First, the fully detailed building models were calibrated into acceptable agreement with the temperature measurements

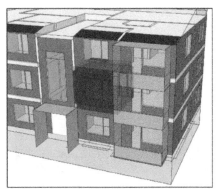
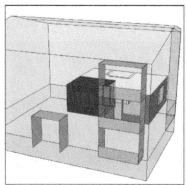
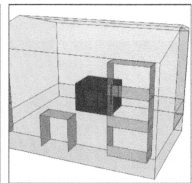

Figure 6. Simulation model detail for different calculation methods: whole building model (left), apartment without neighbouring zones (middle) and single room model (right).

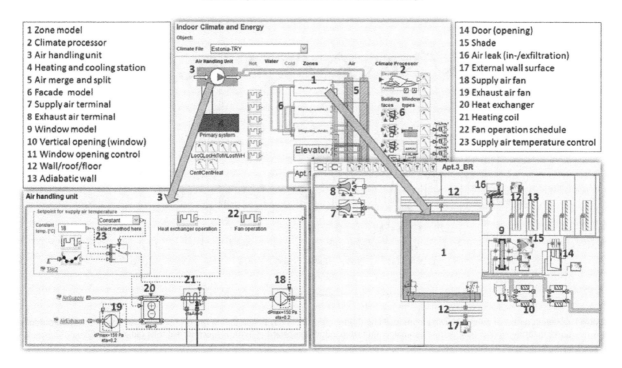

Figure 7. Schematic view of the IDA-ICE environment: example of a whole building model fragment.

from one-month measuring period by changing internal gains, temperature set points for window opening control and by adding internal drapes to the window models. The correlation between the measured and simulated indoor temperature was assessed by linear regression analysis, using Pearson correlation coefficient as one of the indicators.

In order to validate the calibrated models, we used the coefficient of variation of the root mean squared error, CV(RMSE) (4) and the mean bias error (MBE) (5) to quantify the overall accuracy of the simulations (Draper and Smith 1981):

$$CV(RMSE)\,(\%) = \frac{\sqrt{\sum_{i=1}^{n} (Sim_i - Meas_i)^2/n}}{\overline{Meas}} \times 100\%, \quad (4)$$

$$MBE(\%) = \frac{\sum_{i=1}^{n} (Sim_i - Meas_i)/n}{\overline{Meas}} \times 100\%, \quad (5)$$

where $Meas_i$ is the measured value of the variable, Sim_i is the simulated value of the variable, $\overline{Meas}$ is the mean value of the measured variable and n is the number of data points. The CV(RMSE) is essentially the root mean squared error divided by the measured mean of the data (Haberl and Thamilseran 1994). Comparisons were conducted in terms of predicted indoor temperatures. The CV(RMSE) of the hourly simulation results and measured data were calculated (Bou-Saada and Haberl 1995).

To evaluate the quality of the simulation results, additional parameters are used, such as average error percentage (6), average difference between measured and simulated results (7) and average bias (8) for the specified period:

$$Avg.\,Error(\%) = \sum_{i=1}^{n} \left| \frac{Sim_i - Meas_i}{Meas_{Max} - Meas_{Min}} \right| \times \frac{100\%}{n}, \quad (6)$$

$$Avg.\,Dif(K) = \frac{\sum_{i=1}^{n} |Sim_i - Meas_i|}{n}, \quad (7)$$

$$Avg.\,Bias(K) = \frac{\sum_{i=1}^{n} (Sim_i - Meas_i)}{n}. \quad (8)$$

3. Results and discussion

3.1. Model calibration

An example of a model calibration result of a living room for 8-day heat wave period is shown in Figure 8. The simulation results of the calibrated building models for two extreme cases, in terms of DH, are presented in Figure 9. The goal for the validation was to achieve CV(RMSE) values under 5%. It is shown that the validation results show acceptable agreement with the measurements (Figure 10).

3.2. Simulation results for different thermal zoning cases

The simulation time for the largest whole building model, using a high-performance personal computer (housing an Intel© Core™ i7-5820 K processor), was 3 h and 14 min. In comparison, for the apartment-based model with three zones, the simulation time was 1 min and for the single zone model, 8 s.

The calculated simulation evaluation parameters from different thermal zoning methods are shown in Table 2. The average error increases, when simplifications are applied to the whole building models. It can be seen that although CV(RMSE) values in full apartment simulation method and single-zone method remain in similar proportions, the average error is over 10% in four whole apartment model cases and single-zone model cases as well. Results acquired using the single-zone method show mostly lowest agreement with the measurement results, however, being quite close to the apartment-based cases. In four

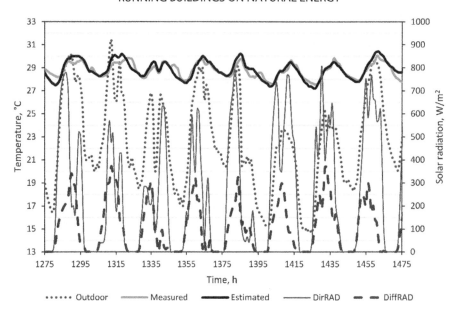

Figure 8. Model validation results of simulated living room (building B4) temperature, 8-day period during the heat wave in 2014 summer. Code: outdoor: ambient temperature; measured: measured indoor temperature; estimated: simulated indoor temperature; DirRAD: direct normal solar radiation; DiffRAD: diffuse solar radiation on horizontal surface.

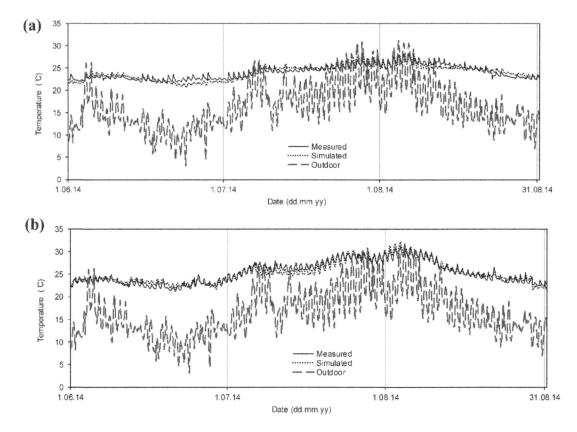

Figure 9. Examples of model validation results: measured and simulated hourly average indoor temperature in selected rooms during the summer period of 1 July to 31 August 2014. Room with the lowest measured temperature excess (DH) – bedroom in building B5 (a) and room with the highest measured DH: living room in building B5 (b).

cases, comparing apartment and single room modelling results, the single room cases give higher DH values, except for the case with building B4, in which the single room method gives lower value. The small change in error values, regarding building B5, could be accounted for the shade casting neighbouring buildings and trees, limiting the effect of direct solar radiation to the building.

Table 3 shows the difference between using standard values for occupant profiles and internal gains according to the methodology (GOV 2012a) and real thermal situation through

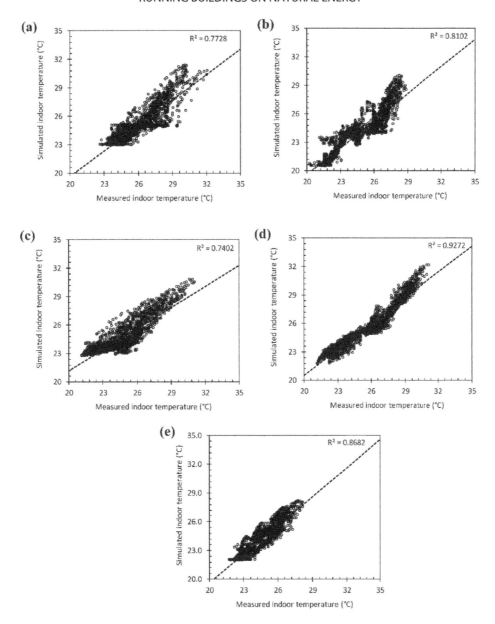

Figure 10. Model validation results: comparison of 2208 h of measured and estimated indoor temperature values: bedroom in building B1 (a), bedroom in building B2 (b), living room in building B3 (c), living room in building B4 (d) and bedroom in building B5 (e).

measurements in the studied rooms. The whole building model and apartment model give mostly lower DH results, as the single-zone method gives higher values for the cases with lower measured DH.

Three of the DH values for single-zone models (B1, B3 and B4), modelled and simulated according to the Estonian methodology (GOV 2012a), are higher compared to the multi-zone model results (Table 4). However, in two cases (B2 and B5), the single-zone model gives lower values. This occurs most likely due to the high temperatures in the neighbouring zones, which is not accounted, in case of the single-zone model, or thermal load shifting due to the movement of the sun and the effect of direct solar irradiation. As the standardized, single-zone simulation results define also the compliance according to the current methodology, it can be seen that rooms, which encountered remarkable overheating in reality, show also non-compliance with the single-zone simulations. The whole building model

and apartment model, however, in building B1, do not indicate non-compliance. The latter case, also when comparing simulations made with climate data from 2014, can be explained with higher internal gains, closed doors between the rooms or lack of window airing in practice, during the measurement period.

3.3. Proposed methodology for measurement-based overheating assessment

Although overheating assessment by calculations for a building in planning stage is required with the state regulations in order to acquire a building permit, the importance of this procedure is often underestimated and calculations are usually done poorly, if at all (Simson, Kurnitski, and Maivel 2016; Tuohy and Murphy 2015). In such cases, it is extremely difficult for the tenant to prove the existence of the problem, as it is only defined as a requirement and a method for evaluating building designs and

Table 2. Evaluation results of the indoor temperature simulations for different modelling detail.

Building no.	Avg. error, (%)	Avg. dif. (K)	Avg. bias (K)	CV(RMSE) (%)	MBE (%)	$DH_{+27°C}$, (Kh)
Measured						
B1	–	–	–	–	–	777
B2	–	–	–	–	–	209
B3	–	–	–	–	–	354
B4	–	–	–	–	–	1053
B5	–	–	–	–	–	35
Calculated: whole building model (fitted)						
B1	7.6	0.8	1.6	4.7	−2.5	765
B2	9.9	0.9	1.3	4.3	−2.1	211
B3	9.6	0.8	1.0	4.1	−0.2	360
B4	5.5	0.6	0.4	1.8	0.1	1065
B5	6.0	0.6	0.5	2.1	−1.5	50
Calculated: apartment model (neighbouring zones removed)						
B1	11.7	1.1	1.7	6.7	−3.7	535
B2	13.7	1.2	2.0	7.9	−3.6	265
B3	10.3	0.9	1.2	5.0	1.9	277
B4	13.9	1.4	3.2	12.4	−4.5	813
B5	6.6	0.6	0.5	2.2	−1.5	29
Calculated: single zone model (neighbouring zones removed)						
B1	12.5	1.1	1.7	6.7	−3.3	641
B2	13.8	1.2	2.0	7.9	−3.2	448
B3	12.1	1.0	1.9	7.6	3.4	936
B4	17.5	1.5	3.6	14.2	−5.1	676
B5	7.6	0.7	0.7	3.0	0.4	230

Table 3. Evaluation of simulated temperature results for different thermal zoning methods using standard values according to the methodology (GOV 2012a) and climate data from summer 2014.

Building, room	B1, bedroom				B2, bedroom				B3, living room				B4, living room				B5, bedroom			
Thermal zoning	MEAS	BLD	APT	SZ	MEAS	BLD	APT	SZ	MEAS	BLD	APT	SZ	MEAS	BLD	APT	SZ	MEAS	BLD	APT	SZ
$DH_{+27°C}$ (Kh)	777	478	535	641	209	190	152	358	354	298	277	936	1053	527	813	676	35	29	39	230
Min temp. (°C)	22.7	23.0	23.0	22.9	20.3	20.4	20.4	20.4	21.2	21.9	21.9	22.0	20.3	20.9	19.2	19.2	21.7	22.0	22.0	21.9
Max temp. (°C)	32.1	31.3	31.7	32.5	28.9	29.9	29.6	30.7	31.1	30.5	30.4	32.7	31.2	30.4	31.7	31.6	28.2	27.9	28.2	30.4
Avg. temp. (°C)	25.8	24.9	24.9	25.0	24.8	23.1	23.0	23.2	24.7	25.2	25.2	25.5	25.3	25.3	24.3	24.1	24.1	23.7	23.7	24.2
Avg. error (%)	–	12.7	9.9	12.6	–	21.0	21.2	20.8	–	10.2	10.5	12.3	–	7.1	10.5	15.0	–	7.0	5.3	7.6
Avg. dif. (K)	–	1.1	1.1	1.1	–	1.8	1.8	1.8	–	0.9	0.9	1.1	–	0.8	1.2	1.3	–	0.6	0.6	0.7
Avg. bias (K)	–	1.8	1.7	1.7	–	4.5	4.6	4.3	–	1.2	1.3	1.9	–	0.9	1.9	2.4	–	0.5	0.5	0.7
Max. dif. (K)	–	4.5	4.5	4.5	–	4.8	4.8	4.3	–	3.5	3.5	4.5	–	3.2	3.4	3.7	–	2.3	2.2	2.7
CV(RMSE) (%)	–	6.8	6.7	6.8	–	18.3	18.5	17.5	–	4.9	5.1	7.7	–	3.7	7.6	9.6	–	2.2	2.1	2.9
MBE (%)	–	−3.8	−3.6	−3.4	–	−6.9	−7.1	−6.3	–	2.0	2.1	3.5	–	−0.1	−4.2	−4.9	–	−1.6	−1.6	0.3

Code: MEAS: measured room; BLD: whole building model; APT: apartment model; SZ: single zone model.

not as an assessment for existing buildings. If the calculations have not been conducted, the acquisition of input data, regarding envelope structures, technical drawings, etc. can be difficult. For such cases, estimating the simulation results, based on real indoor temperature measurements, could act as an efficient and low-cost method.

Different studies have indicated that there is a relatively strong correlation between outdoor and indoor air temperatures at higher ambient temperatures (Nguyen, Schwartz, and Dockery 2014; Walikewitz et al. 2015). For each 1 K increase in outdoor temperature during the warmer periods, the average indoor temperature was found to increase between 0.29 and

Table 4. Evaluation of simulated temperature results for different thermal zoning methods using standard profiles and climate data from TRY.

Building, room	B1 bedroom			B2 bedroom			B3 living room			B4 living room			B5 bedroom		
Thermal zoning	BLD	APT	SZ	BLD	APT	SZ	BLD	APT	SZ	BLD	APT	SZ	BLD	APT	SZ
$DH_{+27°C}$ (Kh)	60	79	189	0	8	0	10	7	55	218	267	319	0	0	0
Min temp. (°C)	22.9	22.9	20.2	19.6	20.4	20.5	22.4	22.5	22.2	24.3	24.2	23.8	21.8	21.8	21.9
Max temp. (°C)	28.7	28.9	30.7	26.6	27.4	26.9	27.7	27.6	28.6	29.0	29.3	29.7	26.4	26.5	26.8
Avg. temp. (°C)	24.7	24.7	23.8	22.3	22.6	22.6	24.9	24.9	25.0	26.2	26.2	26.2	23.5	23.5	24.1
Avg. error, (%)		0.8	13.5		8.2	5.0		0.7	2.9		0.7	1.4		2.1	7.1
Avg. dif. (K)		0.1	1.1		0.7	0.4		0.1	0.2		0.1	0.1		0.2	0.6
Avg. bias (K)		0.0	1.5		0.7	0.4		0.0	0.1		0.0	0.0		0.0	0.5
Max. dif. (K)		0.5	2.8		3.5	3.7		0.3	2.5		0.5	1.3		0.6	1.6
CV(RMSE) (%)		0.0	6.0		3.0	1.8		0.0	0.6		0.0	0.1		0.2	2.0
MBE (%)		0.1	−3.5		1.1	1.3		0.2	0.6		0.2	0.1		−0.7	2.5

Code: BLD: whole building model; APT: apartment model; SZ: single-zone model.

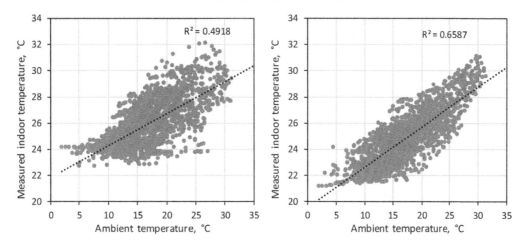

Figure 11. Correlation between measured hourly average indoor temperature and ambient temperature values during the three-month measurement period for two analysed rooms. Lowest linear correlation (left) was found for bedroom in building B1 and highest (right) for living room, building B3.

0.43 K (Tamerius et al. 2013; Nguyen and Dockery 2016). However, it must be stated that besides outdoor temperature, the main factors affecting indoor temperature are solar radiation, internal gains and occupant behaviour (Mavrogianni et al. 2014). Results from our field study are shown in Figure 11, where two measured cases are presented – with lowest and highest correlation of the sample between outdoor and indoor temperature.

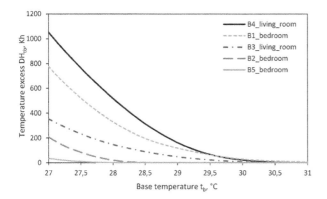

Figure 12. Measured temperature excess (DH) dependence on base temperature t_b in the analysed rooms during the summer period of 2014.

The measured DH for the studied rooms' dependence on the base temperature change is shown in Figure 12.

The simulation input parameters and variables, such as thermal properties of the building, climate data, heat gains and occupant behaviour, are the main source of uncertainty (Encinas and De Herde 2013; Taylor et al. 2014). Thus, the use of a safer, perhaps slightly overestimated approach in overheating assessment can be considered as justified.

The proposed equation, to act as a 'rule of thumb', for correcting the real year base temperature for DH calculations, to make measured room temperature values comparable to standardized calculations, is given as follows:

$$t_{b,n},\text{corr} = t_b + \frac{DH_{t_b,n}}{105}, \qquad (9)$$

where $t_{b,n,\text{corr}}$ is the corrected base temperature for year n, t_b is the base temperature used in standardized calculations, $DH_{t_b,n}$ is the outdoor DH in degree-hours over the base temperature t_b for the measured year and the value 105 is the proposed constant with a reasonable safety factor accounting for the difference in climate data for a real year compared to TRY.

The example using the equation is presented in Figure 13. In the summer of 2014, the correction for base temperature is

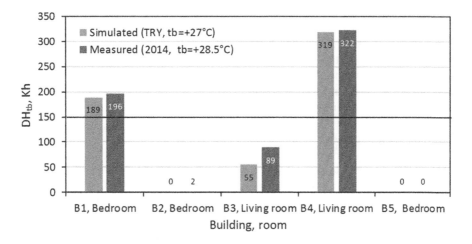

Figure 13. Comparison between measured temperature excess (DH) (summer of 2014) with corrected base temperature $t_b = +28.5°C$ and simulated DH with base temperature $t_b = +27°C$. The 150 Kh line indicating the threshold for compliance.

1.5 K (for $t_b = +27°C$ and $DH_{+27°C,2014} = 157\,Kh$) and the corrected base temperature $t_b = +28.5°C$. For all the cases, the corrected measured values are higher than the regulations-based simulated DH values.

4. Conclusions

In this study, we have compared three thermal zoning methods for summer thermal comfort assessment: two multi-zone approaches, modelling the whole building or apartment, and a single-zone approach, modelling only one room. Simulation results have been evaluated using CV(RMSE), MBE and average percentage error.

The average error increases with the decrease in model detail, thermal connections and air flow routes between neighbouring apartments and rooms. Although in some cases the change in statistical parameters seems low and acceptable in terms of overall indoor temperature prediction, the influence on excess temperature can be substantial, especially in small rooms with large glazing areas.

The analysis of the measurements and simulations reveals that the currently practised single-zone simulation method predicts overheating risk well . In the rooms where overheating was measured, the single-zone model provided the best agreement, indicating that the open doors' assumption of multi-zone model is always not valid in practice. However, as being sensitive for overheating risk estimation, for more accurate predictions, the single-zone method is typically overestimating overheating in the real situation, because it is not accounting the thermal dynamics of the building, heat dissipation between the zones, as well as has limitations in accounting, for example, cross-ventilation. Therefore, the apartment-based multi-zone method gave more realistic results, with little differences to the whole building approach, and can be suggested as an alternative method for more accurate simulations.

It needs to be emphasized that the Estonian single-zone method relies on ventilative cooling through buoyancy-driven window airing, and the fixed window opening position, defined in the regulation, gives sometimes room for interpretations in the design phase and is challenging for simulation tools as well. However, window airing seems to be compensating the oversensitivity of the single-zone model resulting in solid performance according to the measurements of this study.

Although overheating assessment by simulation is required by the state regulations as a precondition to apply building permit, these simulations have been done sometimes poorly. Coupled with the lack of resources, mainly in terms of state officials, and also competence to evaluate the quality and accuracy of the input and output data, the buildings are given permits and are being built with inevitable overheating problems. In such cases, it is extremely difficult for the tenant to prove the existence of the problem as it is defined only as a requirement and method evaluating building designs and not as an assessment for existing buildings. Using the proposed method, it is relatively easy to pre-assess an apartment or living space with only temperature measurements, without having to conduct simulations to prove the existence of overheating problems.

Although the buildings analysed in the current study represent well current construction practice, further research with larger sample representing a larger variety of buildings could be recommended.

Acknowledgements

This research has been conducted within European Intelligent Energy Europe IEE programme project QUALICHeCK: http://qualicheck-platform.eu/ 'Towards Improved Compliance and Quality of the Works for Better Performing Buildings'.

Disclosure statement

No potential conflict of interest was reported by the authors.

Funding

The research was supported by the Estonian Centre of Excellence in Zero Energy and Resource Efficient Smart Buildings and Districts, ZEBE, grant 2014-2020.4.01.15-0016 funded by the European Regional Development Fund, by the Estonian Research Council (Eesti Teadusagentuur) with Institutional research funding grant IUT1−15.

ORCID

Raimo Simson ⓘ http://orcid.org/0000-0001-7271-6081
Jarek Kurnitski ⓘ http://orcid.org/0000-0003-3254-0637

References

Achermann, M. 2000. "Validation of IDA ICE, Version 2.11.06 with IEA Task 12 – Envelope BESTEST." *HLK Engineering*. Hochschule Technik+Architektur Luzern.

Attia, Shady, Jan L. M. Hensen, Liliana Beltrán, and André De Herde. 2012. "Selection Criteria for Building Performance Simulation Tools: Contrasting Architects' and Engineers' Needs." *Journal of Building Performance Simulation* 5 (3): 155–169. doi:10.1080/19401493.2010.549573.

Bateson, A. 2016. "Comparison of CIBSE Thermal Comfort Assessments with SAP Overheating Assessments and Implications for Designers." *Building Services Engineering Research & Technology* 37 (2): 243–251. doi:10.1177/0143624416631133.

Björsell, N., A. Bring, L. I. Eriksson, P. Grozman, M. Lindgren, and P. Sahlin. 1999. "IDA Indoor Climate and Energy." Proceedings of the IBPSA Building Simulation '99 Conference, Kyoto, Japan, September 13–15.

Bou-Saada, Tarek E., and Jeff S. Haberl. 1995. "An Improved Procedure for Developing Calibrated Hourly Simulation Models." Proceedings of the IBPSA Building Simulation '95 Conference, Madison, WI, USA, August 14–16.

Chvatal, K. M. S., and H. Corvacho. 2009. "The Impact of Increasing the Building Envelope Insulation upon the Risk of Overheating in Summer and an Increased Energy Consumption." *Journal of Building Performance Simulation* 2 (4): 267–282. doi:10.1080/19401490903095865.

D3. 2012. *"National Building Code of Finland. Part D3, Energy Management in Buildings – Regulations and Guidelines."* Ministry of the Environment, Department of the Built Environment.

da Silva, P. C., V. Leal, and M. Andersen. 2015. "Occupants' Behaviour in Energy Simulation Tools: Lessons from a Field Monitoring Campaign Regarding Lighting and Shading Control." *Journal of Building Performance Simulation* 8 (5): 338–358.

DECC (Department of Energy & Climate Change) 2014. *SAP 2012: The Government's Standard Assessment Procedure for Energy Rating of Dwellings (2012 ed., revised June 2014).* Watford: Building Research Establishment.

Draper, N. R., and H. Smith. 1981. *Applied regression analysis.* 2nd ed. New York: Wiley.

EMHI. 2015. "Estonian Weather Service." Accessed May 15, 2015. http://www.emhi.ee/?lang = en.

Encinas, F., and A. De Herde. 2013. "Sensitivity Analysis in Building Performance Simulation for Summer Comfort Assessment of Apartments from the Real Estate Market." *Energy and Buildings* 65: 55–65.

ERBD. 2014. "Estonian Registry of Buildings Database." Accessed November 11, 2014. http://www.ehr.ee.

EU. 2010. "Directive 2010/31/EU of the European Parliament and of the Council of 19 May 2010 on the Energy Performance of Buildings (Recast)." *Official Journal of the European Union.* L153: 13–35. doi:10.3000/17252555.L_2010.153.eng

EQUA. 2016. "IDA Indoor Climate and Energy v4.7.1." Equa Simulations AB. http://www.equa.se.

GOV. 2012a. *Estonian Regulation No 63: Methodology for Calculating the Energy Performance of Buildings.* Estonian Ministry of Economic Affairs and Communications.

GOV. 2012b. *Estonian Regulation No 68: Minimum Requirements for Energy Performance.* Estonian Ministry of Economic Affairs and Communications.

Haberl, S., and S. Thamilseran. 1994. "The Great Energy Predictor Shootout II: Measuring Retrofits Savings-Overview and Discussion of Results." In *Energy Systems Laboratory. Technical Report ESL-PA-96/07-03(1).*

Haldi, F., and D. Robinson. 2011. "The Impact of Occupants' Behaviour on Building Energy Demand." *Journal of Building Performance Simulation* 4 (4): 323–338.

Hamdy, M., A. Hasan, and K. Siren. 2011. "Impact of Adaptive Thermal Comfort Criteria on Building Energy Use and Cooling Equipment Size Using a Multi-Objective Optimization Scheme." *Energy and Buildings* 43 (9): 2055–2067.

Heisler, G. M. 1986. "Effects of Individual Trees on the Solar-Radiation Climate of Small Buildings." *Urban Ecology* 9 (3–4): 337–359. doi:10.1016/0304-4009(86)90008-2.

Hilliaho, K., J. Landensivu, and J. Vinha. 2015. "Glazed Space Thermal Simulation with IDA-ICE 4.61 Software-Suitability Analysis with Case Study." *Energy and Buildings* 89: 132–141.

Jenkins, D. P., V. Ingram, S. A. Simpson, and S. Patidar. 2013. "Methods for Assessing Domestic Overheating for Future Building Regulation Compliance." *Energy Policy* 56: 684–692. doi:10.1016/j.enpol.2013.01.030.

Jokisalo, J., and J. Kurnitski. 2007. "Performance of EN ISO 13790 Utilisation Factor Heat Demand Calculation Method in a Cold Climate." *Energy and Buildings* 39 (2): 236–247.

Kalamees, T. 2004. "IDA ICE: The Simulation Tool for Making the Whole Building Energy- and HAM Analysis." Annex 41 MOIST-ENG, Working Meeting 2004, Zürich, Switzerland, May 12–14.

Kalamees, Targo, Simo Ilomets, Roode Liias, Lembi-Merike Raado, Kalle Kuusk, Mikk Maivel, Marko Ründva, et al. 2012. *Construction Condition of Estonian Housing Stock - Apartment Buildings built in 1990–2010: Final Report. In Estonian.* Tallinn: TTU Press.

Kalamees, Targo, and Jarek Kurnitski. 2006. "Estonian Test Reference Year for Energy Calculations." *Proceedings of the Estonian Academy of Sciences Engineering* 12 (1): 40–58.

Kanters, Jouri, Marie-Claude Dubois, and Maria Wall. 2013. "Architects' Design Process in Solar-integrated Architecture in Sweden." *Architectural Science Review* 56 (2): 141–151. doi:10.1080/00038628.2012.681031.

Kropf, Sven, and Gerhard Zweifel. 2002. "Validation of the Building Simulation Program IDA-ICE According to CEN 13791 'Thermal Performance of Buildings – Calculation of Internal Temperatures of a Room in Summer Without Mechanical Cooling - General Criteria and Validation Procedures'." Hochschule Luzern - Technik & Architektur, Luzern.

Maivel, Mikk, Jarek Kurnitski, and Targo Kalamees. 2014. "Field Survey of Overheating Problems in Estonian Apartment Buildings." *Architectural Science Review*, 1–10. doi:10.1080/00038628.2014.970610.

Mavrogianni, A., M. Davies, J. Taylor, Z. Chalabi, P. Biddulph, E. Oikonomou, P. Das, and B. Jones. 2014. "The Impact of Occupancy Patterns, Occupant-Controlled Ventilation and Shading on Indoor Overheating Risk in Domestic Environments." *Building and Environment* 78: 183–198. doi:10.1016/j.buildenv.2014.04.008.

Mavrogianni, A., P. Wilkinson, M. Davies, P. Biddulph, and E. Oikonomou. 2012. "Building Characteristics as Determinants of Propensity to High Indoor Summer Temperatures in London Dwellings." *Building and Environment* 55: 117–130.

Molin, A., P. Rohdin, and B. Moshfegh. 2011. "Investigation of Energy Performance of Newly Built Low-Energy Buildings in Sweden." *Energy and Buildings* 43 (10): 2822–2831.

Nguyen, J. L., and D. W. Dockery. 2016. "Daily Indoor-to-outdoor Temperature and Humidity Relationships: A Sample across Seasons and Diverse Climatic Regions." *International Journal of Biometeorology* 60 (2): 221–229.

Nguyen, J. L., J. Schwartz, and D. W. Dockery. 2014. "The Relationship between Indoor and Outdoor Temperature, Apparent Temperature, Relative Humidity, and Absolute Humidity." *Indoor Air* 24 (1): 103–112.

O'Brien, W., A. Athienitis, and T. Kesik. 2011. "Thermal Zoning and Interzonal Airflow in the Design and Simulation of Solar Houses: A Sensitivity Analysis." *Journal of Building Performance Simulation* 4 (3): 239–256.

ONSET. 2015. "HOBO Temperature Dataloggers, Onset Computer Corporation." http://www.onsetcomp.com/products/data-loggers/u12-012.

Rohdin, P., A. Molin, and B. Moshfegh. 2014. "Experiences from Nine Passive Houses in Sweden – Indoor Thermal Environment and Energy Use." *Building and Environment* 71: 176–185. doi:10.1016/j.buildenv.2013.09.017.

Schweiker, M., F. Haldi, M. Shukuya, and D. Robinson. 2012. "Verification of Stochastic Models of Window Opening Behaviour for Residential Buildings." *Journal of Building Performance Simulation* 5 (1): 55–74.

Simson, Raimo, Jarek Kurnitski, and Mikk Maivel. 2016. "Summer Thermal Comfort: Compliance Assessment and Overheating Prevention in New Apartment Buildings in Estonia." *Journal of Building Performance Simulation* (in press). doi:10.1080/19401493.2016.1248488.

Tamerius, J. D., M. S. Perzanowski, L. M. Acosta, J. S. Jacobson, I. F. Goldstein, J. W. Quinn, A. G. Rundle, and J. Shaman. 2013. "Socioeconomic and Outdoor Meteorological Determinants of Indoor Temperature and Humidity in New York City Dwellings." *Weather, Climate, and Society* 5 (2): 168–179.

Taylor, J., M. Davies, A. Mavrogianni, Z. Chalabi, P. Biddulph, E. Oikonomou, P. Das, and B. Jones. 2014. "The Relative Importance of Input Weather

Data for Indoor Overheating Risk Assessment in Dwellings." *Building and Environment* 76: 81–91. doi:10.1016/j.buildenv.2014.03.010.

Tillson, A. A., T. Oreszczyn, and J. Palmer. 2013. "Assessing Impacts of Summertime Overheating: Some Adaptation Strategies." *Building Research and Information* 41 (6): 652–661. doi:10.1080/09613218.2013.808864.

Tuohy, Paul G., and Gavin B Murphy. 2015. "Are Current Design Processes and Policies Delivering Comfortable Low Carbon Buildings?" *Architectural Science Review* 58 (1): 39–46. doi:10.1080/00038628.2014.975779.

Vuolle, M., and P. Sahlin. 1999. "An NMF Based Model Library for Building Thermal Simulation." Proceedings of the IBPSA Building Simulation '99 Conference, Kyoto, Japan.

Vuolle, M., and P. Sahlin. 2000. "IDA Indoor Climate and Energy – A New Generation Simulation Tool." Proceedings of the Healthy Buildings 2000 Conference, Helsinki University of Technology in Espoo, Finland, 6–10. August.

Walikewitz, Nadine, Britta Jänicke, Marcel Langner, and Wilfried Endlicher. 2015. "Assessment of Indoor Heat Stress Variability in Summer and During Heat Warnings: A Case Study Using the UTCI in Berlin, Germany." *International Journal of Biometeorology*, 1–14. doi:10.1007/s00484-015-1066-y.

Wang, H. D., and Z. Zhai. 2016. "Advances in Building Simulation and Computational Techniques: A Review between 1987 and 2014." *Energy and Buildings* 128: 319–335.

Performance of naturally ventilated buildings in a warm-humid climate: a case study of Golconde Dormitories, South India

Mona Doctor-Pingel, Hugo Lavocat and Nehaa Bhavaraju

ABSTRACT

Golconde, the first modern reinforced concrete building in India, remains as one of the most outstanding examples of climate responsive buildings in the country. This paper will present some of the various passive design strategies employed to ensure thermal comfort without the use of a mechanical cooling system. Among others, the building's surrounding vegetation promoting natural ventilation, its orientation minimizing solar exposure, the ventilated double roof reducing indoor temperatures and the louvres working as solar shading devices are appropriate and efficient strategies for the Indian tropical climate context. The hourly data collected for air temperature, relative humidity and surface temperature over one and half years was used to analyse the impact of those passive strategies on the indoor conditions. This exemplary case study represents a strong case for constructing climate responsive buildings which could address the energy crisis in many countries.

Introduction

India is currently facing a formidable challenge in providing an adequate energy supply to one of the fastest growing energy markets in the world. The building sector's energy consumption, accounting for 35% of the nation's energy use, is growing by 8% annually (Climate Works Foundation 2010). Because of India's rapid economic expansion and urbanization, the floor-space growth in the commercial and residential sector is expected to add the equivalent built area of a new Chicago every year (McKinsey 2010). The increasing reliance on mechanical cooling due to the comfort experience has led to its significant rise over the past few years. This would only lead to the surge in global energy consumption as air conditioners are almost treated as inevitable and are becoming a norm in practise, especially in a tropical climate like India.

Given the increased intensity of energy use, coupled with the changing lifestyles of 1.2 billion people on a quest for an improved quality of life, India must address competence and efficiency in this sector. The coming decade provides an immense opportunity to realize, in practice, the potential for significant energy savings through the construction industry. It would be interesting to study good examples from the modernist movement and check their viability in today's context especially for tropical regions. This could advocate environmental sensitivity as a foundation for the design process.

As a part of the five-year Indo–US joint research program on the energy efficiency of buildings (2012–2017) named 'Centre for Building Energy Research and Development' (CBERD), efforts are being made to provide tangible scientific results that lead to a significant reduction in energy use in buildings in both nations.

This paper will present some of the results of the detailed research being carried out on one of the buildings selected under the CBERD program – Golconde Dormitories of Sri Aurobindo Ashram – a place for disciples and long-term visitors, following the integral yoga of Sri Aurobindo and The Mother (Figure 1).

The building is in an urban setting within the old city precinct (Boulevard area) in Pondicherry, India, (12°N, 80°E) (Figure 2). Designed by Czech-American architect Antonin Raymond, along with George Nakashima and François Sammers – site architects, Golconde is the first reinforced concrete building in India, and is still admired for its outstanding maintenance and thermal performance (Gupta, Mueller, and Samii 2010). The interiors and furniture were specially designed for the building by George Nakashima. By using various passive strategies, this building has been designed to ensure thermal comfort without the use of mechanical ventilation or cooling system. Although built eight decades ago, Golconde remains as an excellent climate responsive building in a warm-humid climate. This is the first time that such detailed study with extensive data logging over a period of one and a half years has been undertaken.

This paper aims to demonstrate that there is a lot to be learnt from a good understanding of the local climate and use of passive design principles which can pave the path to reduced energy demand.

Golconde: site context and design

The building is oriented with its longer axis facing North-South, with a tilt of 20°E, South as shown in Figure 3. This is done as Pondicherry lies on latitude 11 degrees north, thus ensuring

Figure 1. Golconde.

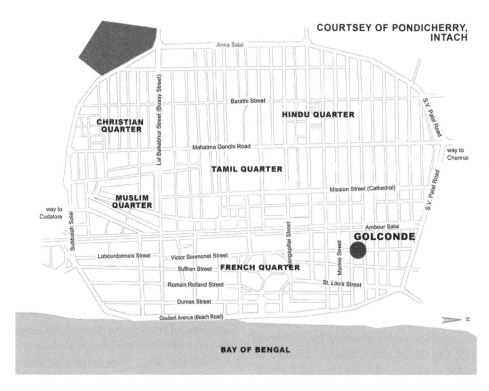

Figure 2. Location of Golconde within the boulevard area, Pondicherry.

alignment with ideal East-West cardinal axis facing due North and South.

The orientation minimizes the solar exposure on the East and West side where the sun is the lowest. It has 51 rooms across 3 floors and a semi-basement. Both the North and South facades are equipped with individually operating horizontal asbestos cement louvres which provide protection from the sun, wind and rain, while allowing for ventilation, ensuring the best combination between natural daylight and solar heat gain. Since it is an occupied building, the louvres are manually operated as per the user's needs, mostly being closed during the day and open during the night.

Access to the rooms is through a continuous corridor along the northern side of the building which also acts as a thermal buffer to the internal spaces. Rooms are separated from the corridor by teak wood sliding doors with staggered slats which

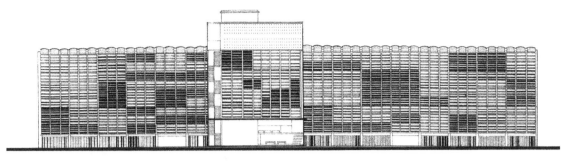

NORTH SIDE ELEVATION

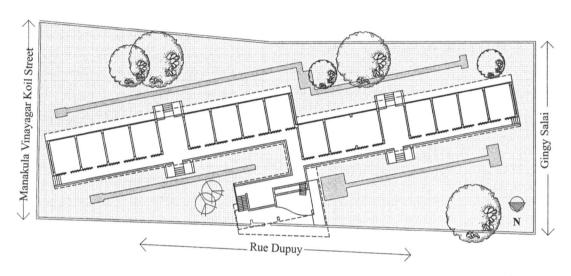

Figure 3. Golconde basement plan and elevation.

Figure 4. Ventilated double roof.

not only allow air to circulate freely when open but also when the doors are closed. The framed reinforced cement concrete (RCC) structure has been left unplastered, while the walls which are made of burnt brick (210 × 100 × 55 mm) have a special 'Chettinad' lime plaster, which is still in its original state. The East and West walls contain little or no openings to reduce the solar heat gain. The floor is made of polished black Cuddapah stones of 63.5 × 63.5 cm with 5 mm butt joints. The carefully landscaped gardens on the North and South side are designed to further

enhance the thermal performance. Golconde's unique ventilated double roof consists of a RCC slab covered with precast concrete shells with an air gap of 10–30 cm in between (Figure 4).

These detailed design interventions keep the insides of Golconde as cool as possible without mechanical ventilation. The comprehensive integration of orientation, structure, interior design and landscaping strategy, adhere to the basic principles

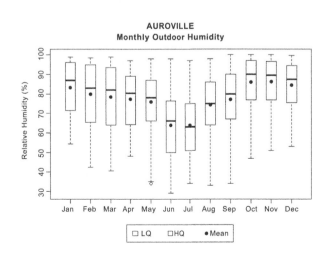

Figure 5. Monthly outdoor air temperature, Auroville Weather station (Komoline) 2014.

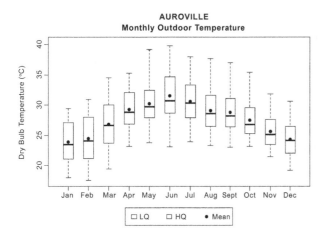

Figure 6. Monthly outdoor Rh, Auroville Weather station, Komoline, 2014.

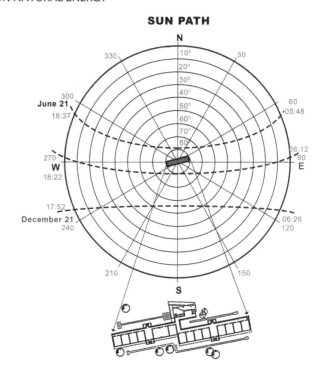

Figure 7. Sun path diagram, Golconde, Pondicherry.

of simplicity without being austere, and having a closeness to nature.

A recent case study carried out in the Reunion Island on a building (ENERPOS) which uses similar strategies such as orientation, landscaping, louvre system, ventilated double roof, also proves the efficacy of these passive strategies in terms of thermal performance (Aurélie et al. 2011).

Climate context

With globalization, our buildings are often standardized and we seem to forget the local constraints and particularities that need to shape them. The bio-climatic concept is one of the pillars of low-energy and sustainable buildings. Consequently, an analysis of the climate context is of prime importance. For this study, the weather station of Auroville, which is the closest to Pondicherry, was used.

The proximity of The Bay of Bengal has a direct impact on the high humidity recorded and the relatively slight temperature fluctuation across the seasons (Figures 5 and 6) (Auroville Weather Station 2015).

As observed in Figures 5 and 6, the temperature in Pondicherry ranges from about 16–40°C throughout the year, with temperatures above 28°C most of the year. There is an average diurnal swing of about 9°C, sometimes going as high as 15°C. The pre-monsoon (July–September) is seen with high humidity and occasional thundershowers and the North-East Monsoon (October–December) season tends to have higher temperatures (averaging 30°C) and higher diurnal swing (averaging 10°C) than post-monsoon and winter (December–February) seasons.

The annual temperature trend curve is opposed to the annual relative humidity (Rh) trend curve (which is a typical trend between temperature and Rh). During June, the hottest month of summer (March–July), temperature peaks reach 40°C. For the same month, the Rh average is 65%. Summer is followed by pre-monsoon and the North-East Monsoon (July–December) and marks a clear change in humidity values due to abundant and regular rainfall. The sun path diagram (Figure 7) shows the sun angles being furthest North during June and furthest South during December (summer and winter solstices).

Golconde – floor plans

Golconde floor plans, logger placement and section are shown in Figures 8 and 9.

Monitoring methodology

A market survey was done and the most appropriate loggers/sensors were identified and procured. Seventeen loggers (with built-in air temperature and humidity sensors) and 17 surface temp sensors were installed in Golconde. To verify the accuracy of the measurement, handheld and logger temperature measurements were cross-checked. Onset HOBO U12-012, Testo410-2 and Extech 30 were used to monitor the air temperature and relative humidity (Figures 8, 9, 10).

Ventilated double roof

Golconde's ventilated double roof consists of an RCC slab covered with precast concrete shell and a ventilated air gap of about 10–30 cm in between (Figure 11).

A large proportion of total heat gain of the building enters through the roof. Efficient roof insulation is crucial to ensure thermal comfort of the room directly below the roof. As shown in Figure 12 (Logger B10 and A7 on 21 June 2014), the efficiency of this ventilated roof allows a reduction by 18°C between the outside and inside surface temperature, occurring under warm and radiant conditions.

Temperature damping (difference between peak outdoor and indoor temperatures) describes the way in which exterior temperatures affect the interiors of a building. As seen from the graph (Figure 12, 21 June 2014), there is a significant

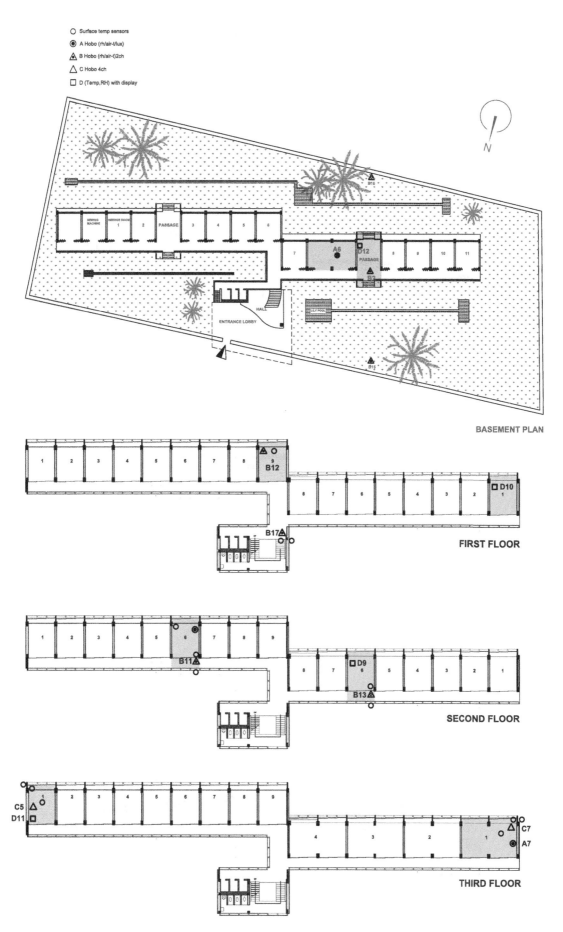

Figure 8. Loggers placement, floor plans, Golconde.

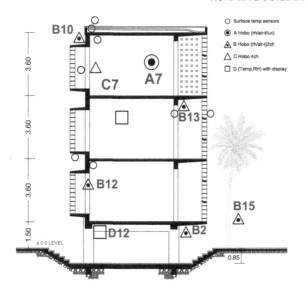

Figure 9. Loggers placement, section, Golconde.

amount of surface temperature damping. However, to get a clearer understanding of the indoor conditions the air temperature damping was also mapped across the year 2014 (Figure 13).

The average maximum damping was found to be around 7°C across the year.

Influence of landscaping

Interestingly in Golconde, the Architect has used a simple landscaping strategy to transform the indoor and the surrounding environmental conditions.

The North garden has been deliberately designed with sparse vegetation, resulting in lighter and dryer air while the South garden has more dense foliage with large tree cover. These tropical evergreen trees of the South garden increase natural shading of the building façade, making the air more dense and moist (Figures 14, 15a and 15b).

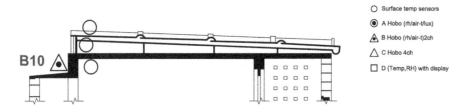

Figure 10. Key for the loggers placement.

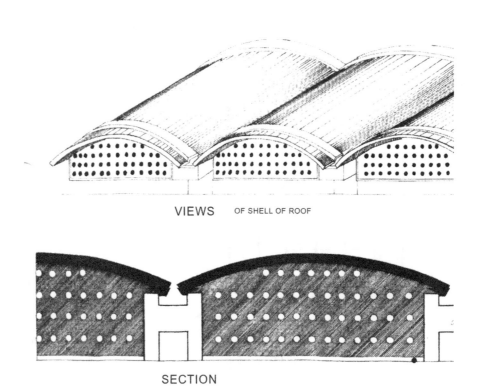

VIEWS OF SHELL OF ROOF

SECTION

Figure 11. Ventilated roof, Louis I Kahn Trophy (2001).

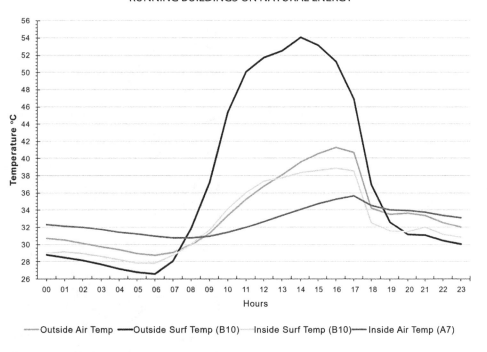

Figure 12. Ventilated roof temperature, 21st June.

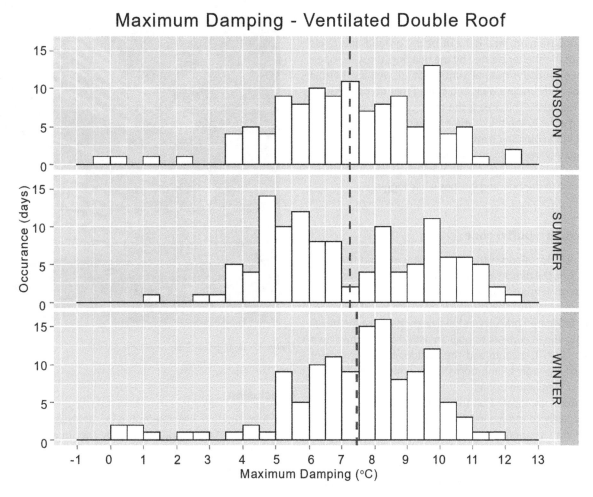

Figure 13. Air temperature damping, 2014.

These gardens are essential to reduce the radiation reflected from the ground (albedo) and prevent hot air from penetrating the building. But more importantly, as shown in Figure 16, during the morning hours, the temperature is lower and the humidity is higher in the South garden than in the North garden (Loggers B16 and B15, 21 June 2014) and vice versa during the noon.

Figure 14. North garden.

This continual temperature difference creates a constant air flow in the building (as the hot air rises making way for the cooler air), especially in the semi-basement passages (Figure 17) making it pleasant to sit even in peak summer afternoons without any mechanical ventilation and thus, most of the common areas (kitchen, dining room, laundry), are in this space.

This semi-basement area is 1.2 m below ground and takes advantage of the shelter created by the building, the wide-open spaces along with the natural ventilation as well as thermal inertia of the surrounding walls and ceiling that do not heat up from solar radiation. This area is the most common space used by the occupants during the daytime, validating that this area is the coolest in the building (Figure 18).

Unfortunately, we could not monitor the wind velocity of the passage, due to limited access to plug points and probability to cause damage to the building finishes while logging.

North corridor – buffer zone

The North corridor, along with the use of louvres, has the function of blocking the direct solar radiation, the rain and to regulate the ventilation (Figure 19). Located on the North and South façade, Golconde's asbestos cement louvres are adjustable. Special, locally fabricated brass levers have been made for the building, allowing the occupants to play an active part in their own thermal comfort. The room and the corridor have permanent air exchange through sliding doors (even when closed) when natural cross ventilation occurs.

This buffer zone located on the northern part of the building is most likely to reduce indoor temperature when the sun is furthest North in summer (Figure 20).

As shown in Figure 20 (Loggers B13 and D9), the room (2W6) has an air temperature of 0.5°C lower than the North passage at the warmest time of the day. With only a small air temperature difference, the air exchange between the room and the corridor is working efficiently. Interestingly, the temperatures inside the room and the corridor are also quite stable with very little fluctuations. The effect of this buffer zone lies in its ability to diminish the northern solar radiation from entering the room.

Figure 15. (a and b) South and facade and garden.

Conclusion

This investigation has attempted an analysis of the thermal performance of Golconde building by continuously monitoring various passive strategies employed. All the graphs have been derived for one day, 21 June 2014, to limit the scope of this paper.

(1) The ventilated double roof acts as a very efficient thermal insulator by reducing about 18°C surface damping from the rooftop to the ceiling surface. On an average, there is about 7°C of air temperature damping recorded throughout the year.

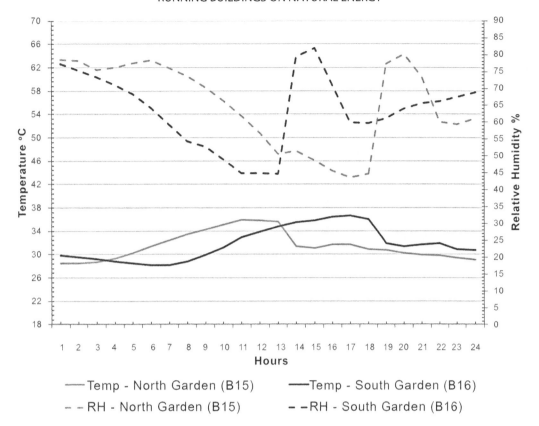

Figure 16. Air temperature and Rh, 21st June.

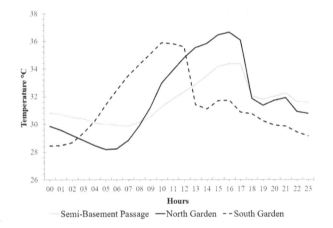

Figure 17. Air temperature – garden and semi-basement, 21st June.

Figure 18. Air temperature – garden and semi-basement, 21st June.

(2) The basement passage has temperatures significantly lower than the gardens (outside temperature) during most of the day, making it the most commonly used space for daily chores and relaxing. This is mainly due to the landscaping strategy in the North and South gardens which create the pressure difference and continuous air movement in the passage.

(3) The corridor in the north which connects the rooms, acts as a very effective buffer zone and helps in reducing the air temperature inside the rooms even during peak summer.

It can thus be concluded that Golconde has an effective passive and natural control system that is responsible for providing a comfortable thermal environment indoors, during the summer.

Scope for further study

This paper presents the first report on the extensive data collected over one and a half years. There is scope for more analysis with respect to performance of the building and individual strategies at different times of the year.

Figure 19. North corridor, Golconde.

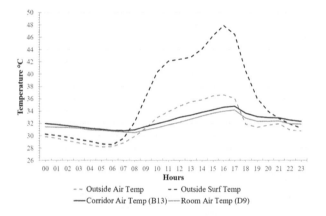

Figure 20. Air temperature in the corridor and room (2W6), 21st June.

Acknowledgements

The authors would like to express our gratitude to the following persons for guidance and support: Dr Brahmanand Mohanty, Dr Chamanlal Gupta (Pondicherry), Prof Rajan Rawal (CEPT, Ahmedabad), Gail Brager (University of California, Berkeley) and Tency Baetens (Auroville Center for Scientific Research).

To use these data anywhere other than the ASR paper, permission from the main author is required.

Funding

The U.S. Department of Energy (DOE) and the Government of India (GOI) provided joint funding for this work under the U.S.–India Partnership to Advance Clean Energy Research (PACE – R) program's 'U.S.–India Joint Center for Building Energy Research and Development' (CBERD) project.

References

Aurélie, L., F. Garde, G. Baird, and M. Franco. 2011. *Environmental Design and Performance of the ENERPOS Building*, Reunion Island, France: The Architectural Science Association.

Auroville Weather Station, Komoline company, 2015. *Auroville Weather Data*. Auroville Weather Station.

Climate Works Foundation. 2010. *Annual Report*. Web. December 2015. http://www.climateworks.org/wp-content/uploads/2014/01/Climate Works-Annual-Report-2010.pdf/.

Gupta, Pankaj Vir, Christine Mueller, and Cyrus Samii. 2010. *Golconde: The Introduction Of Modernism In India*. 1st ed. New Delhi: Urban Crayon Press.

Louis I Kahn Trophy – Golconde: Avant-Garde, NASA (National Association of Students of Architecture, India), 2001.

McKinsey Global Institute. April 2010. *India's Urban Awakening: Building Inclusive Cities, Sustainable Economic Growth*.

Raymond, Antonin, Frank Sammers, and George Nakashima. 2002. *Golconde: A Dormitory for Disciples of Sri Aurobindo, 1936–48*. Pondicherry: Sri Aurobindo Ashram.

The scope of inducing natural air supply via the façade

Peter J. W. van den Engel and Stanley R. Kurvers

ABSTRACT

An overview is given of recent developments in the use of a system of inducing natural air supply via the façade in the Netherlands. This is followed by a review of the results of measurements from climate chamber experiments of its inducing ventilation performance and detailed insights gained from related experiments of climate chamber measurements for a school and a hospital. Finally, lessons learned from practical experience gained in a newly built office and two schools are outlined. These studies of different systems of natural air supply via the facade are used to inform a scoping review of options for use in the design of new buildings using such systems in the future. Because turbulence is an important comfort-parameter, having a positive as well as negative influence on comfort and with physical principles that are, in relation to a number of parameters, still unknown, the issue of turbulence within such systems is discussed in more detail.

1. Introduction

Natural air supply via the façade is one of the most basic and robust options to ventilate buildings. In order to reduce draught problems associated with these natural airflows, a combination of natural supply and mechanical exhaust (ME) might provide a potentially successful option for new buildings. In most cases the use of a ME is necessary in order to guarantee enough fresh air supply via the façade. Due to generally complicated architectural boundary conditions, integration of such a system into the building heat and airflows needs careful attention (van den Engel 2005, 2007). In the future, more fully natural ventilation systems might possibly be used, even in challenging locations, as a result of better integrated design and more physical insight in the character of the natural air flows involved and their control within buildings.

The main research-questions are:

(1) How can draught with natural air supply be prevented? One of the options is high-inducing air supply just underneath the ceiling.
(2) In what way can thermal (dis)comfort due to air flows from natural air supply be evaluated? Recent research (e.g. Ouyang 2006) shows that the method of the EN-ISO 7730 does not give a complete description of the character of turbulence and the effect on thermal comfort.

Between 1990 and 1995 a new air supply system via the façade was developed at Delft University of Technology (DUT) based on a high-speed induced air flow through the external building wall via an outlet below the ceiling indoors (Figure 1; van den Engel

1995a). The main goal of this development is the prevention of draught.

The design principles of the system were subsequently used by architects, the building industry and consultants to develop other similar systems for schools, offices, low energy houses and hospitals. An overview is given of these developments with the performance of several measured and/or simulated situations as well. The focus of this research is on natural air supply and ME. An extensive review of other systems could be the subject of other research. For instance, in the Netherlands there are many different types of air inlets with different qualities and ways of integration. Natural, hybrid and mechanical ventilation could also be compared with each other, as Heiselberg et al. have done (2002). Additional parameters for comparing natural and mechanical air supply systems are:

– Thermal comfort of high-inducing natural air supply can have (almost) the same level as mechanical air supply.
– Air quality of natural air supply is very much dependent on the outdoor quality, but the outdoor quality in Europe is usually much better than the indoor air quality (Ragas et al. 2011). On top of that, the ventilation system itself is a source of pollution, which is often not taken into account (Bluyssen 2009).
– Outside noise till more than 70 dB(A) can be reduced by sound-absorbing air inlets (Schuur 1996). There are many noisy locations where natural air supply is still possible.

Other types of high-inducing ventilation grids that are very much related to the principles and buildings presented here are discussed by Leenaerts and Briggen (2014).

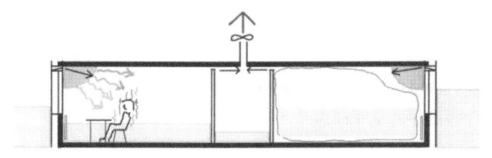

Figure 1. Working principle of the ventilation system with only exhaust in the corridor. Possible pressure differences in and around a building are represented in grey. However, the most effective way of exhaust is direct exhaust of each room. This is the common way of application of the system at the moment.

2. Materials and methods

In the past, ventilation systems were tested and validated using climate chamber measurements that were deemed to provide convincing performance evidence for designers. Currently computational fluid dynamics (CFD) modelling is used extensively in the development of such systems for the early and late stages of the design process. However, the use of these experimental methods cannot completely replicate, or predict, the information gained from actual (comfort) measurements taken in real working or living spaces once the systems are used. That is why the effect in real buildings is and should be evaluated as well.

Additionally, the effect of air turbulence on comfort is not well understood yet and is discussed in more detail below.

3. Theory and calculation

At first, the main physical principles of high-inducing natural air supply are presented. Secondly the advantages of venturi-shaped air inlets are illustrated and finally the effects and characteristics of turbulence are discussed. In fact the presented research consists of two parts: the development and application of high-inducing vents and a discussion about the current evaluation methods of turbulence and draught. A more detailed evaluation is necessary in the future in which the frequency of the pulsating air flow due to turbulence should be taken into account.

Turbulence is not very well understood. In the field of climate design it is usually related to draught. The turbulence intensity is defined as the air velocity divided by the standard deviation of the air velocity. However, the time constant of the measurement is not very well defined; so very short-time and very long-time deviations of the average velocity usually do not play an important role in the comfort-evaluation. On top of that, the frequency distribution of turbulent energy plays an important role and is not evaluated yet. This discussion is relevant for natural as well as mechanical air supply systems. Pressure-controlled fluctuating air flows can also reduce air velocities as seems to be the effect of the Baopt-mechanical ventilation system, which cannot be simulated in CFD yet (Kandzia and Schmidt 2010); so related to turbulence there are still many fields to explore.

3.1. Main principles of high-inducing natural air supply below the ceiling

The physical principles of the high-inducing vents in this system in order to prevent draught are:

(1) High turbulence at the ceiling inlet: high mixing qualities of the incoming air and much induction of surrounding room air into the flow. Limited air flows via air inlets with a small internal height are almost completely mixed before entering the comfort-zone.

(2) In the case of larger air flows, $> 20\,\text{dm}^3/\text{s}\,\text{m}$, the air flow will cling to the ceiling (Coanda-effect) and will not descend and mix with the surrounding air immediately. For larger air flows per metre the Archimedes-number of the inlet, measuring the relationship between the thermal forces and those of velocity, should be lower than circa 0.001 in order to minimize the deflection of the air flow. For instance, an Ar-value of 0.001 is $26\,\text{dm}^3/\text{s}\,\text{m}$ (0°C), supplied with 2.6 m/s via an inlet with a height of 10 mm.

The Archimedes-number (Ar) is calculated as follows:

$$\text{Ar} = \frac{h(T_{\text{room}} - T_{\text{inlet}})g}{U^2 T_{\text{room}}}, \qquad (1)$$

where h is the height of the inlet (m). The temperatures are in Kelvin, g is the acceleration of the gravity: 9.81 m²/s. U is the inlet velocity in m/s.

In the original climate chamber measurements (van den Engel 1995a, 1995b) it was shown that with an air inlet temperature of 0°C, draught rates (DRs) are lower than 20% without extra heating below the air inlet.

The notion of the DR was originally developed for the evaluation of uncomfortable air streams produced by mechanical ventilation systems (Fanger et al. 1988). The DR is defined as the predicted percentage of people that experience discomfort due to draught. This value is measured at critical local points like at the neck height of a sitting person, 1.1 m above the floor. The following equation is used:

$$\text{DR} = (34 - T)(U - 0.05)^{0.62}(3.14 + 0.37UI), \qquad (2)$$

where I is the local turbulence intensity, U the local air velocity and T is the local temperature in °C.

The type of turbulence and airflows that result from natural air supply can have a large influence on the comfort of occupants and depend largely on the design of the inlet-system. The level of the operational temperature in combination with the average local air velocity is the main boundary condition for a positive or negative thermal experience. Turbulence is an additional parameter on air velocity. The level of the turbulence intensity and power spectral density (PSD)-distribution (see Section 3.5) should both be taken into account. For instance, air flows and turbulence via a window with an outside temperature above 15°C are often experienced as positive. Air flows via a fan on the ceiling on a hot day are usually experienced as positive as well. Turbulent cold flows from an air supply device above a working place are often experienced as negative. There is always a strong relation between the temperature, air velocity and feeling of thermal satisfaction. Turbulence is an additional factor that is still underestimated in comfort-evaluation.

In order to evaluate their impact, the use of measurements alone will not give adequate information on how occupants perceive them. Up to now, high-inducing air inlets in the façade have not been systematically studied with subjective evaluations of occupants in a room.

3.2. Venturi shape of an air inlet

The lowest pressure differences within such a system are possible with an inlet shape for a venturi flow. By using that form it is already possible to produce an airflow that clings to the ceiling (> 1 m) at a pressure difference below 3 Pa. Moreover, the same venturi flow that is used for daytime ventilation could be modified for use in night-time purging of heat from the mass of a building. In this way the amount of façade-elements such as windows or grilles for passive buildings can be reduced (Figure 2).

The required comfort level can also be sub-optimally realized with air inlets that are already available within the Dutch building industry and if used as such the most influential design attribute of the system is the actual size and position of the inlet within the room. Minor additions and adaptations to proprietary systems are necessary in order to place the inlet at the right location (Figure 7(c)) or to preheat the air with inserted simple heating pipes (Figure 8). A disadvantage of the common solutions is that the total pressure difference over commonly available inlets is rather high (> 5 Pa) due to the non-optimization of the system integration into the building.

The maximum velocity via an opening in a venturi can be calculated with the following equation:

$$U = \left(\frac{2\Delta P}{\rho} \right), \tag{3}$$

which can also be derived from Bernouilli's law.

3.3. Origin of the DR approach

One of the first studies on the effect of airflows on human sensation responses was a laboratory study in the mid-1980s where 100 subjects were exposed to a turbulent airflow (Fanger and Christensen 1986). The study was performed in a laboratory space, which had one outside wall and no windows. The airflow was coming from behind and aimed at the position of the seated subjects. The subjects participated in three-, two- and a half-hour experiments and the supplied air velocities fluctuated in a random manner. The turbulent airflow was characterized by the mean velocity and the turbulence intensity, defined by the standard deviation divided by the mean velocity. Figure 3 shows the found relationships between the mean air velocity, air temperature and the percentage of dissatisfied at a turbulence intensity of 34.6%.

It appeared that people are more sensitive to draught from a turbulent air flow, coming from a mechanical ventilation system, then from a laminar flow. The back of the head was the most draught-sensitive part of the body. No significant differences between the draught sensitivity of men and women were found. Also it was found that there were substantial inter-individual differences in draught sensitivity.

In a later study the impact of turbulence intensity on the sensation of draught was investigated with more levels of turbulence (Fanger et al. 1988). Fifty subjects were exposed to air flow with low (Tu < 12%), medium (20% < Tu < 35%) and high (Tu > 55%) turbulence intensity and to six mean air velocities between 0.05 and 0.40 m/s. The sensation of draught was significantly influenced by the turbulence intensity. A model was presented which predicts the percentage of people dissatisfied due to the perception of draught as a function of air temperature, mean velocity and turbulence intensity (Figure 4). The time constant of the velocity measurement equipment was 0.1 s; so fluctuations up to 10 Hz could be measured.

These laboratory studies form the basis of the DR equation of the international standard EN-ISO-7730. The aim of this standard is to provide a calculation method of the allowed values

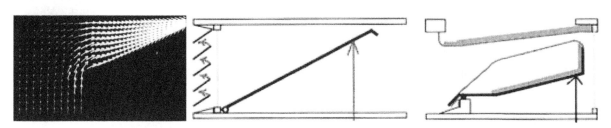

Figure 2. CFD-simulation of airflow in a venturi-shaped inlet (left) and two types of similar air inlets (right) that have been tested in a climate chamber. In the right model sound-absorbing materials are added. However, a venture-shape in itself is not favourable for sound-isolation. A protecting element at the outside (like simulated in the CFD-calculation) could improve this. This way of protection does not have an effect on the air resistance.

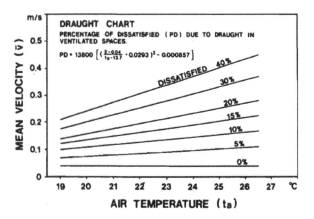

Figure 3. The draught chart. The chart applies to sedentary persons, wearing normal indoor clothing, exposed to air velocity in the occupied zone of ventilated spaces. The turbulence intensity is 34.6% (Fanger and Christensen 1986).

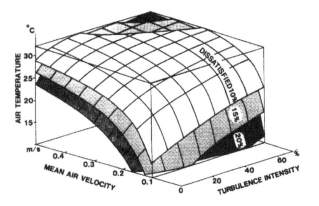

Figure 4. The draught risk model. The surfaces shown correspond to 10%, 15% and 20% dissatisfied, respectively. The axes represent turbulence intensity, mean air velocity and air temperature (Fanger et al. 1988).

of the combination of air temperature, air velocity and turbulence intensity. Given the nature of the laboratory studies, where airflows were generated mechanically, this equation to determine thermal comfort due to allowable air velocities cannot be considered valid for airflows with a natural characteristic, like airflows through windows and grids. Because turbulence in air flows through façade inlets can be very different, it is better to check, not only the turbulence intensity, but also the effect of the PSD on comfort (see Section 3.5).

3.4. Air flows and the human skin

Wind or airflows can be characterized in terms of time and space. Temporarily fluctuations can be hourly, diurnal or seasonal. Spatially, wind can be characterized by two horizontal vectors, the azimuth and a vertical vector, the elevation. Air does not flow exactly horizontally; the elevations varies with speed. In the laboratory studies of Fanger and Christensen the airflows were coming from behind the subjects; however recent studies show that people are more sensitive of airflows coming from the front (Simone and Olesen 2013). Conclusions from experiments carried out in controlled environments should be treated carefully because human perceptions might not be similar to the field observations (Djamila, Ming, and Kumaresan 2014). Various studies (e.g. Humphreys, Nicol, and Roaf 2016) show that

people adapt themselves to the environment and control the environment to suit their comfort preferences depending on the physical and social context. What might be considered an unpleasant draught in one situation may be a welcome breeze in another.

When we perceive a warm or cool environment, we do not actually sense the temperature of the air or heaters directly, but rather we notify a change in conditions by our nerve endings, the thermoreceptors, which send signals in certain frequencies to the hypothalamus in the brain (Zhang 2003). The combined effect of air temperature and air movement on the human skin reaches the brain coded as frequency of nerve signals. Research shows that the frequency of change in airflows of between 0.2 and 1.0 Hz had a strong cooling effect on subjects (Huang, Ouyang, and Zhu 2012). Fluctuations with frequency between 0.3 and 0.5 Hz seem to have the greatest impact and are most likely to be perceived as draught (Madsen 1984). Higher frequencies (between 0.7 and 1.0 Hz) could also significantly affect thermal comfort (Ouyang 2006). Humans sense air temperature and air movement differently in cold and hot environments because the skin includes more cold than warm receptors (de Dear 2010).

3.5. Power spectral density

As turbulence intensity describes turbulence as a function of time, turbulence can also be studied in the frequency domain. PSD shows how power is distributed over a range of frequencies. The velocity, measured by time, can be transformed by Fourier transform into a PSD graph. In Figure 5 an example of the PSD is given. The PSD $E(f)$ is defined as follows (Ouyang 2006):

$$E_f \propto 1/f^\beta. \tag{4}$$

To be able to quantify the power spectral characteristics of airflow, the so-called β-value can be used, which represents the negative slope of the logarithmic power spectrum graph. The higher the value, the higher the power in the low frequencies, which represents large period eddies (Ouyang 2006). Figure 6

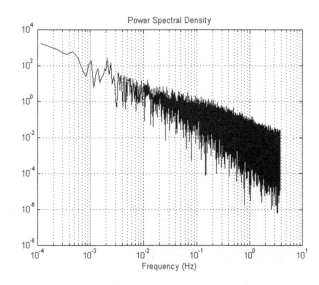

Figure 5. PSD (Djamila, Ming, and Kumaresan 2014). On the vertical axis the power spectrum density $E(f)$ is presented. The sampling interval is 0.13 s.

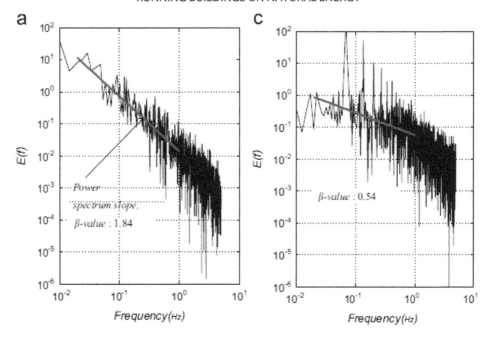

Figure 6. Representation of power distribution and β-value for natural (left) and mechanical airflow (right). The sampling interval is 0.1 s. The hot-wire anemometer could measure fluctuations up to 5 Hz (Kang, Song, and Shiavon 2013).

gives examples of β-values of natural and mechanical airflows (Kang, Song, and Shiavon 2013).

When we take a closer look at the difference in spectrum characteristics between natural and mechanical airflows, we see that natural airflows have higher β-value than mechanical ones.

When the airflow velocity is reduced, an increase in β-value of mechanical airflows is noticed. This is also seen to a lesser extent for natural airflows, although the β-value remains higher than 1.1 (Ouyang 2006).

Concerning human perception, it was found that airflows with higher β-values are perceived as more pleasant. The optimum range for β-values is between 1.41 and 1.80 and are mainly found in natural airflows (Kang, Song, and Shiavon 2013). This supports the findings that people prefer airflows from systems with the characteristics of those found with natural ventilation.

Much of the research on air velocities and advanced data-processing and analyses techniques is dependent in part on the sensitivity of the measuring devices. New research should clearly specify the required response time of the devices used, including conventional anemometers or modern ultrasonic measurement equipment. In order to develop more insight in the complete spectrum of PSD, high sensitive measurement equipment is recommended, but it is not clear yet how much effect high frequencies have. For many years it is already possible to measure fluctuations up to 100 Hz (Mayer 1987; van den Engel 1995a).

3.6. Still or variable?

The current air velocity standards seem to apply to situations with mechanical ventilation under (mainly winter) circumstances where they could cause draught complaints, but in situations with natural ventilation, higher temperatures and/or personal control opportunities other airflow characteristics might be preferred.

In warmer climates or under summer conditions draught might not be the main concern, but sufficient air movement is a significant advantage (Aynsley 2012). In naturally ventilated environments, airflows play an important role in controlling indoor air quality and thermal comfort. In extensive field studies (Arens et al. 2009; Hoyt, Zhang, and Arens 2009) it is found that people who feel cool prefer less air movement and those who feel warm prefer more air movement, even though the occupants did not have much control over air movement. For people voting for thermal sensation values of 0.7–1.5, the air movement limit can be extended to 0.8 m/s. Standards such as EN-ISO-7730 were based on the concept of avoiding any disturbing or undesirable air movement (draught). More recent standards, such as ASHRAE 2010 and EN-NEN-15251, appreciate that higher air speeds are useful in offsetting increased temperatures and allow higher airspeeds to maintain both the same total heat transfer from the skin and to promote thermal comfort. Another additional effect of air movement, especially around the face, is an increase of perceived air quality, even though the measured air quality was the same (Zhang et al. 2010). This could have important consequences for saving energy through the use of elevated air motion by means of personal control systems, operable windows and (advanced ceiling) fans (Huang et al. 2013).

The concept of 'alliesthesia' is another way of assessing air velocities and this knowledge can be used for designing naturally ventilated environments. When a certain environmental stimulus is able to restore the internal state of the body to its set-point, this is perceived as pleasant, this can be called positive alliesthesia. A stimulus that enlarges the discrepancy between the internal state and its set-point will be perceived as unpleasant: negative alliesthesia (de Dear 2010). Examples of negative alliesthesia are heating sources in a warm environment and temperature gradients that cause 'warm head' and 'cold feet'. Positive alliesthesia can be a fire place in a cold room and 'cool head' and 'warm feet'. A slight breeze can bring thermal pleasure when the core temperature is slightly above neutral, but the same air movement can be perceived as an unwanted 'draught' if the core temperature is below its set-point.

It seems that in steady-state conditions, people can be comfortable, but to be very comfortable, the environment should rather offer sensational transience, asymmetries and personal control (Arens et al. 2010; de Dear 2010; Brager, Zhang, and Arens 2015). In terms of design this asks, for instance, for new ideas on window design, the use of shadings, operable windows and ventilation grids in relation to summer and winter conditions.

4. Applications

The results of four different categories of functions are discussed in which decentralized natural air supply is a realistic option: offices, schools, hospitals and houses.

For a school (Regionaal Opleidingencentrum (regional educational centre [ROC] of Twente)) and a hospital, detailed climate chamber measurements have been carried out. Another school (De Schakel in Utrecht) has been measured just before being in use (Versteeg 2014). All the measurements are related to or based on the comfort-criteria of the EN-ISO 7730.

The office (Amolf Research Centre in Amsterdam), schools and houses are also built; so there is user experience as well. Up to now this is positive; however, no systematic user evaluation has been carried out yet.

All the presented systems have natural air supply and ME.

4.1. Offices

One of the requirements of clients can be as much natural ventilation as possible, as was the case for the occupants of the three storey Amolf (Figure 7(a)). Here an induced natural supply system in the façade was combined with operable windows, a concrete core activation (CCA) system and a convector placed at floor level underneath the air inlet. Moreover, an outside sunshade was added. The pressure difference via the façade is lower than 10 Pa. Each room has a separate ME. Up to now, no draught problems of the occupants have been reported.

4.2. Schools

An inducing natural air supply can be a low energy option for the successful ventilation of classrooms compared with more electricity hungry mechanically driven, ducted ventilation systems with heat recovery (MVHR). This is largely due to the fact that classrooms often have a high occupancy level and consequently a high internal heat load. Not only does the induced natural ventilation provide fresh air at a far lower mechanical energy cost, but the use of the body heat to warm the air passing through the class room and into the rest of the building will also limit the necessary amount of heating required for comfort in the school as a whole. On top of that, such systems offer much free cooling potential.

The starting point of the design should be provision of natural fresh air with draught prevention resulting from effective flow control. With heated pipes (Figure 8), air can be warmed up to above 0°C to produce a DR lower than 20% (Cauberg-Huygen 2005). A combination is possible of induced natural ventilation

with radiators below the inlet or CCA. For large schools with CCA built over several storeys, it is essential that the system is automatically closed off at the end of the occupancy period. This is necessary in order to reduce energy loss and to keep the temperature of the CCA as low as possible and to prevent a too large temperature difference between class rooms.

For schools a Dutch guide has been developed to improve the climatic conditions (ISSO 2008). In this publication, mechanical ventilation systems are compared with systems with natural air supply. In order to prevent draught, principles of induced air near the ceiling are presented. This Dutch guide has been a source of both information and inspiration for architects and consultants as to how to improve the quality of natural air supply while preventing down draughts and minimizing energy consumption.

A comparable, but more compact, system with only one heated pipe in the inlet and a radiator against the façade has been applied for the school De Schakel (Versteeg 2014). CFD-simulations and thermal comfort measurements according to EN-ISO 7730 in this other school confirmed the earlier conclusion that the DRs with this ventilation system are low and below 20%.

4.3. Hospitals

Hospitals are special environments with vulnerable occupants who may suffer from ventilation-related health problems. Natural air supply is an interesting alternative for mechanical supply, as there might be smaller risk of contamination by pathogenic micro-organisms than can be found in the air supplied by handling units and supply ducts with filters that are regularly, or irregularly, maintained and changed (Green 2011).

To achieve very low DRs, air should be preheated by more than 0°C, depending on outdoor conditions. Measurements of patient rooms for a new hospital indicate that DRs below 15% are possible with Archimedes numbers lower than 0.0003 (10 mm inlet size, 15°C, 2.6 m/s, 26 dm^3/s m or 10°C, 3.6 m/s, 36 dm^3/s m). These inlet temperatures are comparable with those produced by mechanical air supply systems (Peutz 2008). In this case, two air inlets with a width of 0.84 m are used. The limitations of the width of the air inlet, and the channelling of the flow from outside to inside (screens inside the plenum) also led to DR reductions (Figure 9).

4.4. Houses

Disappointing monitored results from houses with MVHR in the Netherlands led to a growing interest in induced natural air supply systems via the façade. After health problems in the district Vathorst of the city of Amersfoort in 2007, MVHR seemed to be the main cause. However, this was mainly due to the poor installation of the MVHR system and lack of operable windows. Nevertheless, additional research initiated by the Ministry of Health made clear that the appreciation of natural air supply systems by occupants is still higher than MVHR (RIVM 2011).

Natural air supply with ME and full MVHR systems are all being continuously improved and both have a future, despite some recent adverse press on the subject. The choice of an optimal system often depends on the outdoor and architectural contexts,

a

b

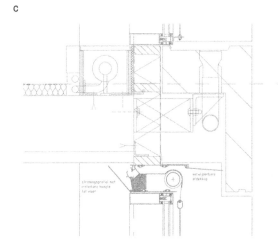

Velocity
1.930E+00
1.818E+00
1.706E+00
1.594E+00
1.482E+00

1.258E+00
1.146E+00
1.034E+00
9.220E-01
8.100E-01
6.980E-01
5.860E-01
4.740E-01
3.620E-01
2.500E-01

c

Figure 7. (a) Top figure: overview of the façade with outside sunshade. (b) Result of a CFD-simulation on an exceptional very cold winter day. The window is 13°C. There is a convector (540 W) underneath the air supply. Floor (24°C) and ceiling (30°C) are heated as well (CCA). The presented air velocities are between 0.25 and 1.93 m/s (dark area near the ceiling). However, CFD tends to overestimate the throw of the jet with small air flows (van den Engel 1995a). In the occupancy zone the air velocities are lower than 0.25 m/s. The supplied air temperature is −10°C (far below zero). 70 m^3/h is supplied via an inlet with a height of 7 mm and width of 2 m (Ar = 0.0036). (c) An architectural detail of the natural air inlet-system for the Amolf Research Centre Architect: Dick van Gameren. The inlet is hidden behind a concrete façade-element.

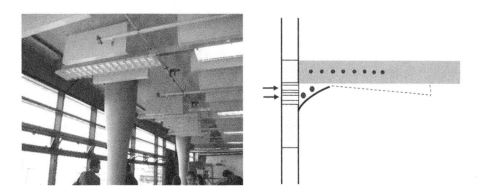

Figure 8. Air inlet-system of the ROC of Twente. Air inlets are placed in the façade, air is preheated with pipes and supplied to the room via a narrow slit near the ceiling.

Figure 9. Air inlet-system for a hospital in a mocked-up experimental room.

With an effective control of fresh air supply via the façade, natural supply with ME is now an option as well. An inlet with an air flow lower than 10 l/s/m, located more than 1.80 m above the floor, will not produce a serious draught problem. The higher the location above the floor, the lower the draught risk. For houses the average pressure difference is much lower due to a small required air flow per metre façade: around 1–2 Pa. An example of an air inlet near the ceiling of a (near) zero energy house is the Brabant House air inlet-system of Renz Pijnenborgh (Buma 2014 and Figure 10). This solution was developed by the architect because there is no radiator or convector underneath the air inlet, which increases draught risk. The house is heated by internal 'warm walls'.

5. Conclusions

(1) In order to optimize occupant comfort and save energy, induced natural ventilation (air supply via the façade and exhaust via shafts) is a challenging option. One of the most practical solutions for the prevention of discomfort in dwellings or spaces with low occupancy is to ensure that an amount of air can be supplied via the façade of around 10 l/s m, from outlets below the ceiling. With this limited amount of fresh air supply, mixing of cold and warm surrounding air is possible within a zone of circa 1 m from the façade. When larger airflows are required, such as in classrooms, the pressure differences should be increased, in order to prevent draughts and to direct and maintain the air flow as close to the ceiling possible. Induced natural ventilation works well in a hybrid combination with a ME system that can guarantee the required pressure difference (5–10 Pa) and direction of flow. In that case it is possible to realize sufficient air supply and an air flow that clings to the ceiling.

including conditions of ambient noise, pollution and the height of the spaces above ground. Often reported or recorded ventilation complaints are noise from the fans from ME and MVHR systems (van Dijken and Boerstra 2011) and high measured levels of CO_2 in the sleeping rooms. The reported problems could be solved, for instance, by lower air velocities in the ventilation system (reduction of noise), local CO_2-controlled valves or more conscious usage of the opening of the vents (reducing CO_2-level). However, CO_2 is not a good indicator of perceived air quality (Sassi 2016). There is no evidence to suggest yet that a technically more advanced system, such as a CO_2-controlled system, leads to more occupant satisfaction or a more robust system. Moreover, occupant complaints or user satisfaction are not always directly related to physically measurable parameters. Probably not all dominant physical parameters are known, and psychological parameters, such as experience and expectation, may have a great impact as well.

More than half of the houses in the Netherlands are equipped with natural supply systems. For a passive or zero energy house, MVHR recovery used to be the standard for the past decade or so.

(2) For inducing vents for an outer wall, a wide range of natural air supply systems, principles and products are available for use in carefully designed systems to reduce draught risk and

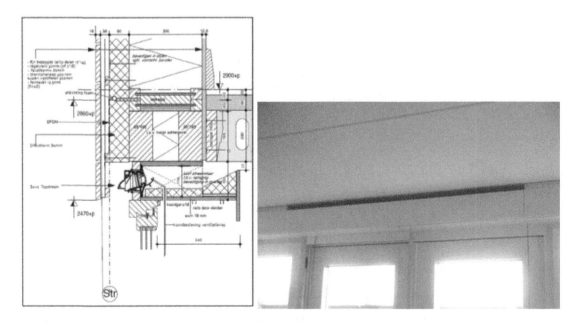

Figure 10. Example of air supply near the ceiling with the option of a curtain-location that will not disturb the air flow that enters the room (Renz Pijnenborgh).

energy used for ventilation. A further development of these systems can involve the simplification of the shape of these vents in order to reduce the pressure loss below 5 Pa and make them more appropriate for night ventilation.

(3) Induced natural air supply is also an option when the outdoor temperature rises and night ventilation alone is not enough to fulfil the comfort requirements. In that case a combination of the system with cooled floors and/or ceilings is possible (Figures 7 and 8).

(4) Induced natural air supply systems are often used in conjunction with operable windows for summer-time comfort cooling as well. More insight into the physical principles of air flows, and turbulence, is necessary to reinforce the current trend in induced natural ventilation systems becoming a more commonly specified solution.

(5) Points to watch out for in the design of such systems are the prevention of return air flows and the reduction of resistance of incoming air across the external wall. Heat losses via the air inlet should be limited as much as possible through the use of good detailing to eliminate cold bridging in the systems in their passage through the external envelope of the building.

(6) To assess the effectiveness of induced natural air supply systems, it is necessary to also evaluate their performance using real occupants in their ordinary work or living spaces in order to give more insight into the actual comfort conditions they create.

(7) The measurement of the PSD in summer and winter, combined with occupant feedback, will give more detailed insight in the thermal quality of air flows across the systems.

(8) Evaluations of the energy consumption of natural air supply systems in use are also necessary to gauge the real-world cost benefits of their adoption in real buildings.

Disclosure statement

No potential conflict of interest was reported by the authors.

References

Arens, E., M. A. Humphreys, R. de Dear, and H. Zhang. 2010. "Are 'Class A' Temperature Requirements Realistic or Desirable?" *Building and Environment* 45: 4–10.

Arens, E. A., S. Turner, H. Zang, and G. Paliaga. 2009. "Moving Air for Comfort." *ASHRAE Journal* 51 (5): 18–21.

Aynsley, R. 2012. "How Much Do You Need to Know to Effectively Utilise Large Ceiling Fans?" *Architectural Science Review* 55 (2): 16–25.

Bluyssen, P. M. 2009. *The Indoor Environment Handbook – How to Make Buildings Healthy and Comfortable.* London: Taylor & Francis.

Brager, G., H. Zhang, and E. Arens. 2015. "Evolving Opportunities for Providing Thermal Comfort." *Building Research & Information* 43 (3): 274–287.

Buma, W. 2014. "De Brabantwoning." *TVVL-magazine* 2014 (4): 34–36.

Cauberg-Huygen. 2005. *ROC van Twente te Hengelo. Klimaatkameronderzoek.* Report 2004.2297-1v2.

de Dear, R. 2010. "Thermal Comfort in Natural Ventilation – A Neurophysiological Hypothesis." Proceedings of conference: adapting to change: new thinking on comfort cumberland lodge, Windsor, April 9–11, 2010. London: Network for Comfort and Energy Use in Buildings.

van Dijken, F. A., and A. C. Boerstra. 2011. "Onderzoek naar de kwaliteit van ventilatiesystemen in nieuwbouweengezinswoningen."

Djamila, H., C. C. Ming, and S. Kumaresan. 2014. "Exploring the Dynamic Aspect of Natural Air Flow on Occupants Thermal Perception and Comfort." Proceedings of 8th Windsor conference: Counting the cost of comfort in a changing world cumberland lodge, Windsor, April 10–13, 2014. London: Network for Comfort and Energy Use in Buildings.

van den Engel, P. J. W. 1995a. *"Inducing Vents and Their Effect on Air Flow Patterns, Thermal Comfort and Air Quality." Indoor Air, an Integrated Approach, 1994.* Gold Coast, Australia: Elsevier Science.

van den Engel, P. J. W. 1995b. "Thermisch Comfort en Ventilatie-Efficiency Doort Inducerende Ventilatie via de Gevel." Thesis, TU-Delft.

van den Engel, P. J. W. 2005. "Healty Climate in Schools Due to Ventilation an Slab Heating." Proceedings of Clima 2005, Lausanne.

van den Engel, P. J. W. 2007. "Typologies of Hybrid Ventilation in Schools." Proceedings of Clima 2007, Helsinki.

Fanger, P. O., and N. K. Christensen. 1986. "Perception of Draught in Ventilated Spaces." *Ergonomics* 29 (2): 215–235.

Fanger, P. O., A. K. Melikov, H. Hanzawa, and J. Ring. 1988. "Air Turbulence and Sensation of Draught." *Energy and Buildings* 12 (1): 21–39.

Green, J. 2011. http://www.ted.com/talks/jessica_green_are_we_filtering_the_wrong_microbes.html.

Heiselberg, P., A. van der Aa, S. Aggerholm, Å. Blomsterberg, M. Citterio, W. de Gids, Y. Li, et al. 2002. "Principles of Hybrid Ventilation." Summary of IEA-ECBCS Annex 35 "hybrid ventilation in new and retrofitted office buildings," Aalborg University.

Hoyt, T., H. Zhang, and E. Arens. 2009. "Draft or Breeze? Preferences for Air Movement in Office Buildings and Schools from the ASHRAE Database." Proceedings of healthy buildings, Syracuse, NY, September 13–17.

Huang, L., Q. Ouyang, and Y. Zhu. 2012. "Perceptible Airflow Fluctuation Frequency and Human Thermal Response." *Building and Environment* 54: 14–19.

Huang, L., Q. Ouyang, Y. Zhu, and L. Jiang. 2013. "A Study about the Demand for Air Movement in Warm Environment." *Building and Environment* 61: 27–33.

Humphreys, M., F. Nicol, and S. Roaf. 2016. *Adaptive Thermal Comfort. Foundations and Analysis.* London: Routledge.

ISSO. ISSO-publicatie 89. 2008. "Binnenklimaat scholen."

Kandzia, C., and M. Schmidt. 2010. "Bauer Entzaubert? Untersuchung zu instationären Betriebsweiseneines Luftfürungssystem." RWTH-research, Technik 13.

Kang, K., D. Song, and S. Shiavon. 2013. "Correlations in Thermal Comfort and Natural Wind." *Journal of Thermal Biology* 38: 419–426.

Leenaerts, C. L. M., and P. Briggen. 2014. "Innovatief ventilatieconcept voor 'frisse school' MFA het nest. Ventilatieconcept met natuurlijke toevoer voorkomt klachten binnenklimaat." *Bouwfysica* 3: 2–6.

Madsen, T. L. 1984. "Why Low Air Velocities May Cause Thermal Discomfort." Proceedings of indoor air, Stockholm.

Mayer, E. 1987. "Draught Measurements in Ventilated and Non-ventilated Buildings." Proceedings of the 8th AIVC-conference, Überlingen.

Ouyang, Q. 2006. "Study on Dynamic Characteristics of Natural and Mechanical Wind in Built Environment Using Spectral Analysis." *Building and Environment* 41: 418–426.

Peutz. 2008. "Eindrapportage Klimaatkameronderzoek Reinier de Graaf." Report BZ 469-4.

Ragas, A. M. J., R. Oldenkamp, N. L. Preeker, J. Wernicke, and U. Schlinck. 2011. "Cumulative Risk Assessment of Chemical Exposures in Urban Environments." *Environment International* 37: 872–881.

RIVM. 2011. "Kwaliteit van mechanische ventilatiesystemen in nieuwbouw eengezinswoningen en bewonersklachten." Report 630789006.

Sassi, P. 2016. "Evaluation of Indoor Environment in Super-insulated Naturally Ventilated Housing in the South of the United Kingdom." Proceedings of the 9th Windsor conference, Windsor.

Schuur, A. 1996. "Ventilatie, geluidsbelasting en gevelontwerp." Ventilatie via de gevel, een delicate balans tussen luchtkwaliteit en energiegebruik. Symposiumbook DUT.

Simone, A., and B. W. Olesen. 2013. "Preferred Air Velocity on Local Cooling Effect of Desk Fans in Warm Environment." Proceedings of 3th AIVC–4th TightVent–2nd venticool joint conference, Athens, September 25–26.

Versteeg, H. 2014. "Natuurlijke Toevoer van Ventilatielucht in Onderwijsruimten." *TVVL-Magazine* 2014 (4): 2–4.

Zhang, H. 2003. "Human thermal sensation and comfort in transient and non-uniform thermal environment." Thesis, UC Berkeley.

Zhang, H., E. A. Arens, D. Kim, E. Buchberger, F. S. Bauman, and C. Huizenga. 2010. "Comfort, Perceived Air Quality, and Work Performance in a Low-power Task-ambient Conditioning System." *Building and Environment* 45 (1): 29–39.

The importance of air movement in warmer temperatures: a novel SET* house case study

John J. Shiel ⓘ, Richard Aynsley, Behdad Moghtaderi and Adrian Page

ABSTRACT

Surface temperatures increased rapidly in the last 100 years by 1 K (Kelvin), and could increase by a further 1.4 K in just 35 years, challenging building designers to provide comfort while minimizing carbon emissions. Ways to do this are with more effective indoor ventilation and lighter clothing for high temperatures and humidity, but some thermal simulation and rating systems do not consider these aspects. This paper reports on a novel simulation case study that estimated the heating and cooling energy used in a home in a Warm Temperate climate under a changing climate with a case-study method that used (1) CSIRO's Climate Futures online tool; (2) the Australian Nationwide House Energy Ratings Scheme (NatHERS) AccuRate simulation and ratings tool; (3) a special CSIRO humidity research engine and (4) alternative Standard Effective Temperature (SET*) comfort approaches. The results showed that (A) one SET* approach with air movement, changed clothing and occupant acclimatization saved over 95% of the NatHERS residential heating and cooling energy, and should be included in NatHERS; and (B) residential retrofits or occupant education is needed for warming temperatures.

Introduction

Australia has an international obligation to reduce its carbon emissions to near zero by 2050 (ASBEC 2016, 22) and the building sector accounts for around one-quarter of carbon emissions (ASBEC 2008, 8). Over half of those emissions come from Australia's existing residential stock, which has a very poor thermal performance of less than 3 stars of the maximum rating of 10 of its Nationwide House Energy Ratings Scheme (NatHERS) (Ren, Chen, and Wang 2011, 2400). Furthermore, the ownership of residential air conditioning was 40% with a low growth rate around 1994, but is now almost universal with an ownership rate of 94% in 2012 (DEWHA 2008a, 48; Ryan and Pavia 2016, 9–6). This rapid growth of air conditioners coincided with the introduction of performance regulations and the adoption of NatHERS in the early 1990s.

Many researchers have also reported differences between the results of NatHERS simulations of homes and monitored temperatures, energy used or occupant thermal comfort (Saman et al. 2008; Williamson, Soebarto, and Radford 2010; Dewsbury 2011; Kordjamshidi 2011; Page et al. 2011; Copper 2012; Ambrose et al. 2013; Saman, NCCARF, and UniSA 2013; Moore et al. 2016).

From a ventilation perspective, NatHERS underestimates the cooling effect of air movement at high humidity levels (Shiel et al. 2014; Daniel et al. 2015). Global warming and rising energy prices provide the impetus to see if air movement can provide the key to affordable comfort and reduce NatHERS energy use and carbon emissions.

The goal of this study is to understand the interactions between internal building ventilation, comfort and energy use as it relates to existing housing. The heating and cooling energy of a case study was simulated with two additional comfort approaches that could provide greater benefit of air movement, for two climate change scenarios in 2050.

Energy used by Australian buildings

Figure 1 shows the annual energy per square metre for all buildings in Australia in 2011 by category, against the annual energy consumed for those categories (BZE 2013; ABS 2013a; ABS 2013b), and this highlights the building categories requiring retrofit attention since it shows:

- Building categories consuming large amounts of energy with the 'All Residential' category at 400PJ consuming around twice as much energy as all other categories put together, and correlating well with other estimates (DEWHA 2008a, 20; Ryan and Pavia 2016, 9–5).
- Building categories with high energy intensities (per square metre) higher on the vertical axis, where energy is consumed greatly for some building categories, although they may contain fewer buildings for example, 'Museums and Galleries', 'Hospitals', 'Aged Care' and 'Accommodation'.

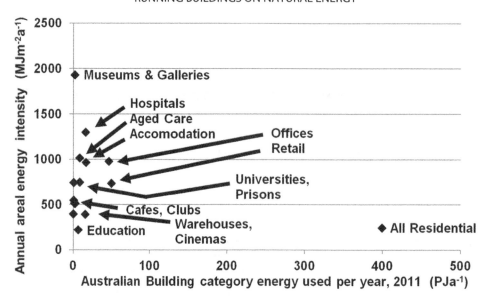

Figure 1. Average annual energy intensity by area (MJ/a.m²) for the total energy used per year (PJ/a), for Australian building categories (Source: Shiel with data from BZE 2013; ABS 2013a, 2013b).

So, the energy used in all residential buildings is very large, and around 40% of that is consumed for heating and cooling (DEWHA 2008a, 24).

Climate change in Australia

Australian surface temperatures have already warmed by around 1.0 K (Kelvin) from the 1910 to 2010 average (CSIRO and BoM 2015, 42), and are projected to rise by 1.4 K between 2010 and 2050 according to the mean temperature increase of an extreme Representative Concentration Pathway (RCP), RCP8.5 (CSIRO and BoM 2015, 92). A RCP is the current climate scenario approach for future GHG concentrations (not emissions) that was adopted by the Intergovernmental Panel on Climate Change for its fifth Assessment Report (AR5) in 2014 (IPCC-AR5-WGI 2013, 19), and supersedes the Special Report on Emissions Scenarios (SRES) approach. While these temperature increases are not predictions, the GHG concentration of the planet has historically been tracking close to the temperatures projected by the extreme pathway RCP8.5 (IPCC-AR5-WGI 2013, 104).

Temperatures for some Warm Temperate climate cities like Sydney in the East Coast South sub-cluster are projected to rise by around 1.8 K for the 50th percentile of RCP8.5 from 1995 to 2050, which would mean that Sydney temperatures may become more like those of Sub-tropical Brisbane (BoM 2016) if RCP8.5 eventuated, although for a proper comparison, humidity should be considered as well. Sydney has recently broken historical temperature records including with extreme overnight temperatures of 30°C (Blumer and Mayers 2017).

Occupant behaviour for comfort

Occupant behaviour often can have a very significant influence that can be positive or negative on the energy used to maintain comfort in a building (DEWHA 2008b, 25, 135–144; White 2009; Candido 2011; Janda 2011).

When applying the term 'passive' to a building, there could be two dimensions considered to maintain thermal comfort (White 2009; Candido 2011):

- The degree of passive solar design of the buildings, that minimizes the need for energy-consuming mechanical operations,
- The degree to which the occupants, consciously or subconsciously, actively improve their comfort on a regular basis for example, by
  - Controlling vents and ceiling, extractor, wall or personal fans;
  - Operating doors, windows or window coverings and shades such as curtains, blinds and shutters;
  - Changing clothing and metabolic levels and
  - Moving closer to windows for radiative solar warmth in cold weather, or away from the windows in summer.

The bottom left segment of Figure 2 shows the ideal situation where as little energy as possible is consumed in a building

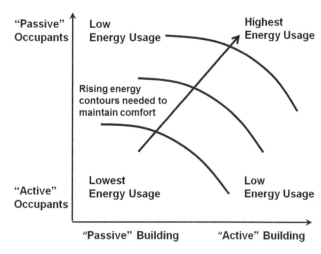

Figure 2. Indicative contours of energy needed to maintain occupant comfort, for 'Active' and 'Passive' occupants and buildings (Source: Shiel).

that has a passive solar design (a 'Passive' building), and the occupants actively manage their comfort ('Active' occupants) before resorting to using minimum mechanical energy. An 'Active' building could be defined as one where much mechanical energy is primarily used for comfort. 'Passive' Occupants could be defined as those who do not actively manage personal or building controls to maintain comfort.

Towards a zero carbon 2050

The vast majority of Australia's existing dwellings will last another 40 years, but their carbon emissions are too high for the zero carbon transition. This highlights the importance of retrofitting Australian houses or occupant behaviour to adapt to a potential step-change increase in temperature in the coming decades, and to reduce energy and carbon emissions at the same time.

Thermal comfort for building occupants

Thermal comfort is difficult to measure since it is highly subjective, but one widely accepted definition is:

> ... that condition of mind that expresses satisfaction with the thermal environment. (ASHRAE-55 2010, 11)

The main thermal comfort approaches to the reporting and evaluation of comfort as experienced and predicted for building occupants are:

- The Effective Temperature (ET) index and its variants, including the Standard Effective Temperature (SET*) index
- The Predictive Mean Vote (PMV)
- The adaptive thermal comfort (ATC) index

ET and variants including SET*

The ET scale was developed by Houghton and Yagloglou in 1923, and includes the impacts of temperature and some thermal sensation humidity effects on perceived comfort in a standard environment, with revisions culminating in the New Effective Temperature (ET*) suitable for many environments in 1974 by Gagge (Parsons 1993; Szokolay 2004, 21). ET* is defined as the temperature of a standard environment where the relative humidity is set at 50%, mean radiant temperature (MRT) equals air temperature and air speed, v, is set at 0.15 m/s. It accounts for the comfort experiences of persons with sedentary activity and light clothing (Parsons 1993, 212–213).

The initial SET comfort approach for a standard environment was developed by Gagge, Nishi and Gonzalez in 1972 as an extension to ET* (Gagge, Fobelets, and Berglund 1986; Fountain and Huizenga 1995; ASHRAE-55 2010). It was further extended into the SET* suitable for many environments from the experiments of Kansas State University on human temperature, clothing and activity level that led to a 'two-node model' of the human core and skin (Aynsley 2008, 10). This took into account skin set-points and the body's thermoregulatory physiology that regulate skin blood flow, sweat rate and increases of metabolic heat by the shivering of muscles, and modelled heat interactions from the human core due to metabolic activity with respiration and

the skin's convection, radiation and evaporation due to wettedness. Skin wettedness is the degree to which the skin is covered in sweat and this has a direct correlation with comfort sensation under warm conditions and assists in determining the physiological limit to heat stress, which has important implications for climate change. The SET standard environment definition from Parsons is

> the temperature of an isothermal environment with air temperature equal to mean radiant temperature, 50 per cent relative humidity, and still air ($v < 0.15\,\text{ms}^{-1}$) in which a person with a standard level of clothing insulation would have the same heat loss at the same mean skin temperature and the same skin wettedness as he does in the actual environment and clothing insulation under consideration ... The ET* is therefore equivalent to the SET for sedentary activities (and light clothing). (Parsons 1993, 213) NB: Parson's variable 'v' refers to air velocity

Predictive Mean Vote

Fanger's PMV (Fanger 1972) evolved from climate chamber and laboratory tests and surveys. The predicted percentage of occupants dissatisfied gives a method of calculating comfortable static indoor temperatures for the air-conditioning industry to use in buildings.

There were moves to include humidity in PMV by using ET* to replace operative temperature in the PMV equation and call it PMV*, but Fanger felt that humidity was not a major concern around neutrality conditions (Parsons 1993, 216).

Adaptive thermal comfort

The ATC index is based on the theory that people fundamentally adapt to the climates and temperatures of the spaces they occupy and if they are not comfortable, they alter either themselves or their surroundings to return to comfort. It is a robust algorithm from which comfort temperatures for locally adapted populations can be predicted and was produced from the strong correlation between outdoor temperatures and reported indoor temperatures based on occupant surveys undertaken during field studies in real buildings. The algorithm finds a 'neutral temperature' which uses the running mean of the outdoor air temperature (with several versions that use the mean of temperatures calculated from 7 days up to a maximum of 30 days) to reflect changing thermal conditions and for which occupants feel no discomfort. It assumes an acceptable indoor air temperature within a 7 K range, depending on the circumstances and preferences of individuals within particular buildings in a single climate (Auliciems et al. 1997; de Dear et al. 1998; Nicol, Humphreys, and Roaf 2012; Humphreys, Nicol, and Roaf 2015).

The ATC index is traditionally seen as being applicable for people in naturally conditioned, or free running, buildings with little or no exposure to air conditioning (ASHRAE-55 2010, 23), with all types of clothing for prevailing mean outdoor temperatures between 10°C and 33.5°C. Any exposure to air conditioning would decrease the climatic adaption of the occupants and their perceived level of thermal comfort while occupying the naturally conditioned building (Romm 2016). Users of a school assembly hall designed with the ATC approach in Rio de Janeiro after similar buildings in rural regions of Brazil were well received complained that the hall was 'too hot'. An hour-by-hour survey conducted of complainants' location over a period of weeks revealed that they spent 77% of their time in air-conditioned spaces, causing them to lose their local climate adaption (Grimme, Laar, and Moore 2003).

The most comprehensive comfort approach

The PMV excludes humidity and is for air-conditioned buildings, while the ATC index is valid from 10°C and 33.5°C. The ET* does not consider evaporative heat loss when the operative temperature exceeds 30°C (Aynsley 2008, 10) and so the most comprehensive thermal index is SET*, since it depends on all six basic parameters air dry-bulb temperature (Ta); MRT; humidity; air speed (v); the clothing level (CLO) and metabolic rate (MET); is applicable over the complete range of cold to hot conditions (Parsons 1993, 214–215; Aynsley 2008, 9); and caters for evaporative heat loss at high temperatures. SET* was difficult to calculate when it was first formulated, but the algorithm can be easily implemented now on present-day computers, with programmes readily available including online (Fountain and Huizenga 1995).

NatHERS

Australia's Nationwide House Energy Rating Scheme (NatHERS) is a thermal performance ratings system that calculates the annual required energy per square metre of a dwelling to remain comfortable. It gives a location-dependent rating from 0 stars (a poor dwelling) to 10 stars (one requiring little energy for comfort) for the building fabric's thermal performance.

There are three accredited NatHERS software applications that allow assessors to describe the house thermal characteristics such as geometry, materials, zones (i.e. enclosed spaces such as rooms) and location. These generate a file that is used by the NatHERS CHENATH engine which uses a frequency–response thermal model and a multi-zone airflow model (Walsh and Delsante 1983; Ren and Chen 2010) to perform the temperature, energy and star rating calculations (Chen 2008; Baharun, Ooi, and Chen 2009; Chen and CSIRO 2016).

NatHERS comfort

The comfort approach of NatHERS ('ET*90%') combines:

- The ET* to calculate the effective temperature;
- the ATC to define heating and cooling set-points for Australian locations based on Auliciem's neutral temperatures with the narrow 90% acceptability of temperature comfort band for occupants and
- Szokolay's cooling algorithm for air movement at 50% humidity (Aynsley and Szokolay 1998).

NatHERS underestimates the cooling effect of air movement at high humidities due to using simplifying assumptions in the CHENATH engine (Aynsley 2012); Szokolay's 50% humidity model for the cooling effect of air movement; and a simplified psychrometric chart approach (Aynsley and Szokolay 1998; Delsante 2005).

Energy and star rating calculations

The NatHERS CHENATH engine calculates the annual hourly zone temperature, humidity and air movement for each zone for the year, driven by the reference meteorological year (RMY) weather file and zone type heating loads to determine any space heating and cooling energy requirements of the conditioned zone types. It begins the heating of a zone when the environmental temperature is below the heating set-point at the end of any hour.

For cooling, CHENATH calculates the hourly zone temperature, humidity and air velocity calculations for an extended comfort region of a psychrometric chart, including for any available natural and mechanical ventilation (Delsante 2005). If the zone hourly temperature still rises above the cooling set-point for the location (Ren, Wang, and Chen 2014), then the cooling system is switched on. The energy to keep within the comfort temperature bands is summed for all zones for the year, divided by the conditioned area and adjusted for the area and climate to correct for a bias against small homes, and this adjusted required energy per square metre is used to find the NatHERS star for each location.

Air movement

CHENATH did not take into account wind direction, opening sizes and location in the façade and between zones, stack effect, nor comfort from air ventilation, until an air network model and the Szokolay cooling effect of air movement was included in 2002 (Delsante 2005). Since then, it has incorporated assumptions of occupant behaviour and site characteristics including that the home is occupied continually for ventilation and conditioning control operation; the local dwelling's wind speed, direction and gusting are often based on nearby open terrain airports (Willrath 1998, 3.26; Ren and Chen 2015) or coastal weather stations that can measure wind at 10 or 16 m heights; and that well-sealed dwellings are energy-efficient and healthy.

However, improvements continue to be added to CHENATH regarding large permanent openings, cross-ventilation and an improved stack effect (Ren and Chen 2010; Chen 2014). Also, a more sophisticated natural ventilation model has now been implemented in the 2014 CHENATH engine (Ren and Chen 2010), although it still suffers from the historical nearby airport issues. Recently, there has been more research underway regarding infiltration, particularly with regard to existing dwellings (Ren and Chen 2015), and the third generation of NatHERS tools is currently being developed (Byford, Hage, and CRC-LCL 2016).

Case study

A simulation study was conducted on a small house in Adelaide, South Australia, currently in a Mediterranean climate (Köppen climate classification Csa) as well as for two warmer climate scenarios in 2050. The purpose was to estimate the effect of air movement on the predicted NatHERS required energy savings for comfort with alternative comfort approaches.

House details

Figure 3 shows the small 77 m^2 one-bedroom case-study house, with (a) the 3-dimensional view of the house; (b) the Plan;

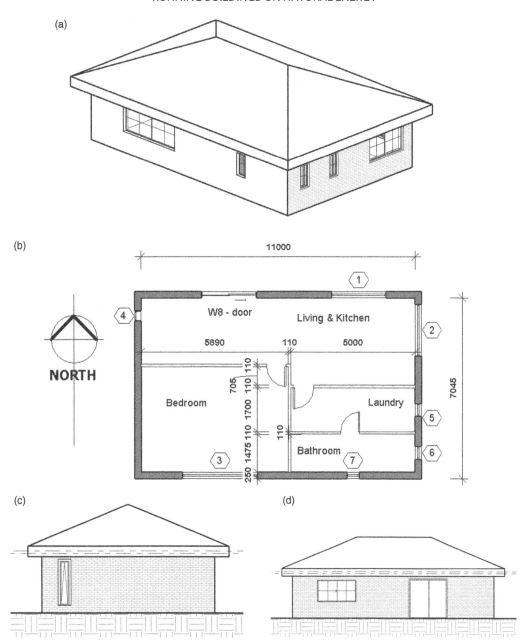

Figure 3. Adelaide case-study house. (a) 3-d view. (b) Plan. (c) West elevation. (d) North elevation (Source: Shiel).

(c) the West elevation and (d) the North elevation. The house had brick veneer walls, a concrete slab on ground floor and metal roof, a construction type typical of modern houses in the region. The case-study walls had an outer skin of 11 cm single brick, an inner load-bearing timber frame with R2.5 K m² W⁻¹ and reflective insulation, and were lined with plasterboard. There were aluminium single-glazed windows and 60 cm wide eaves; R1 K m² W⁻¹ edge insulation for the concrete slab; R4 K m² W⁻¹ insulated ceilings; a light-coloured highly ventilated metal roof and exposed internal single brick walls between the Kitchen/Living Room and the Bedroom, and between the Kitchen/Living Room and the Laundry.

Infiltration and ventilation assumptions

Aluminium sliding windows and glass door had insect screens and a small gap size into which paper would not fit. So, the house was designed to be the highest sealed level by AccuRate standards.

The Bedroom and Living/Kitchen had a 140 cm diameter ceiling fan, with all rooms except the Bedroom having a sealed exhaust fan. The long wall was facing due North and no study was carried out on the wind data in AccuRate, but the NatHERS protocol of modelling neighbours and fences was followed which affected natural ventilation assumptions.

Method

Two SET* comfort approaches were researched:

- One with a narrow comfort band corresponding to the ET* standard environment similar to that deployed in NatHERS (and denoted here 'SET*90%') and with air speeds $v = 0.1$ and 1.5 m/s and CLO = 0.8, and

- Another with a wider comfort band and no restrictions on the six parameters with more air speeds and CLO values (denoted here 'SET*80%').

The comfort band for SET*80% came from Parsons '"Slightly cool … " to Slightly warm … ' sensation ranges of 17.5–30°C (Parsons 1993, 215) since his 'neutrality' range of 22.2–25.6°C was judged to be too narrow, and in the absence of field studies with SET* data. The wider range which was called an 80% acceptability range for this study was adopted for SET*80% because ASHRAE recommends the 80% comfort band for offices for the adaptive comfort model, rather than the more restrictive 90% band (ASHRAE-55 2010), and for home occupants:

- They are exposed to the outside temperatures more than office workers by opening windows and going outside (Willrath 1998, 3.17);
- They occupy zones with more adaptive controls than in offices for example, in bathrooms and bedrooms (Peeters et al. 2009);
- Their comfort expectations may be different (Henriksen 2005, 3.16; Peeters et al. 2009) and

- Each comfort sensation vote is significant with few occupants in the dwelling compared to an office and so there could be wider variability.

Figure 4 illustrates the overall concept of the method, where the grey discomfort region of the alternative SET*80% comfort approach is identified, and how this is transformed into a black discomfort region for which NatHERS can calculate the conditioned energy. The transformation is made by mapping the SET*80% thermostats into corresponding naturally conditioned temperatures.

Figure 5 shows a diagram of the steps involved in this innovative method to estimate the house SET* heating and cooling energy, for 1990 and the climate change scenarios. The steps were:

(1) Determining the weather input files for 2050 from climate change modelling (top right of Figure 5);
(2) Determining the conservative new climate 2050 thermostat settings for the naturally conditioned temperatures corresponding to the SET* approach (bottom centre of Figure 5) and

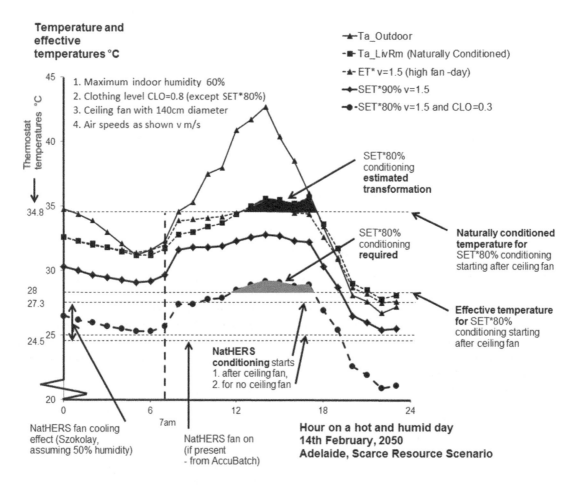

Figure 4. The method concept, where the grey discomfort region of the alternative SET*80% comfort approach is identified, and how this is transformed into a black discomfort region for which NatHERS can calculate the conditioned energy by mapping the SET*80% thermostats into corresponding free running or naturally conditioned temperatures (Source: Shiel).

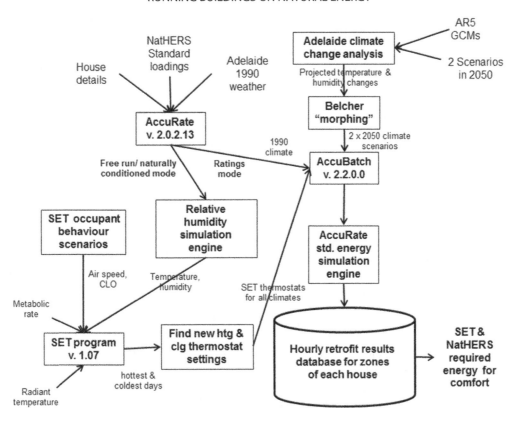

Figure 5. Method for transforming the SET* thermostats into naturally conditioned temperatures to estimate the house SET* heating and cooling energy for three climates (Source: Shiel).

(3) Calculating the energy and star ratings (bottom right of Figure 5) where the heating and cooling is only switched on when the temperatures are outside the SET* thermostat settings (see Table 1).

(1) Weather input files for 2050

The hourly weather data for two new climate scenarios for the year 2050 were developed from (A) the changes of the monthly temperature and humidity from 1995 to 2050 for two scenarios; (B) the modification of these monthly temperature and humidity parameters using Belcher's 'morphing' technique to eliminate bias (Belcher, Hacker, and Powell 2005) and (C) these hourly 'morphed' results were added to the NatHERS Adelaide 1990 RMY weather file. This gave a downscaling method which is

popular with building researchers (Hacker et al. 2008; Ren, Chen, and Wang 2011; Chen, Wang, and Ren 2012). It should be noted that the 1990 and 1995 base years are considered close enough when modelling climate change over 50-year epochs.

The input files were found for two climate change scenarios in 2050 (with more details provided in the Discussion section) for:

- An Extreme Climate Change scenario corresponding to RCP8.5, and
- A Scarce Resource scenario corresponding to RCP4.5.

The Australian Climate Futures (ACF) online tool from the Commonwealth Scientific and Industrial Research Organisation (CSIRO 2015) was used to find the monthly mean temperature and rainfall (for humidity) parameters, which provided

Table 1. Thermostat settings (°C) for Adelaide house case study, for 1990 and the 2050 Extreme Climate Change scenario

Comfort Criteria	For heating			For cooling	
Zone	Living/ Kitchen	Bedroom		Living/ Kitchen	Bedroom
Period	07:00–24:00	01:00–07:00	08:00–09:00, 16:00–24:00	07:00–24:00	16:00–09:00
NatHERS ET 90% acceptability (1990)	20	15	18	25	25
SET*90% relative humidity (%)				40	40
SET*90% naturally conditioned AccuRate thermostat temperature (2050)	16.7	14.6	16.8	27.3	25.1
SET*80% acceptability	17.5	17.5	17.5	30	TBD
SET*80% assumed settings (2050)	18	16	18	28	26
SET*80% relative humidity (%)				38	40
SET*80% naturally conditioned AccuRate thermostat temperature (2050)	15	13.7	14.2	34.8	30.4

Source: Shiel, with SET*80% acceptability from (Parsons, 1993, 215) (TBD – see Method section)

sufficient accuracy to project the heating and cooling energy requirements for 2050 for a Warm Temperate climate (Chen, Wang, and Ren 2012, 232). The ACF ranked the HadGEM2-ES general circulation model as one of the most suitable for both scenarios for 2050 for the Adelaide region with high confidence (Shiel et al. 2017). Only the 'Maximum Consensus' climate futures were considered since the aim of the study is to compare among adaptation techniques rather than explore the impacts of other climate futures.

(2)   New climate 2050 thermostat settings

There were three steps to estimate the thermostat settings of the two new 2050 climates. Firstly, a naturally conditioned or free-running analysis was performed to determine the internal temperatures and humidities, using a CSIRO research relative humidity CHENATH simulation engine for AccuRate v.2.0.2.13 (D. Chen. 2015 "RE: Modified AccuRate engine with internal Ta, RH and P." Personal Correspondence, February 27).

Secondly, SET* and ET* indices were calculated for each zone of the case study using ASHRAE's Thermal Comfort Tool v1.07 (Fountain and Huizenga 1995) with the six inputs: the AccuRate naturally conditioned (1) temperatures and (2) humidity, and assuming (3) that the MET was sedentary that is, MET = 1; (4) that the MRT was the same as the naturally conditioned temperature; (5) that the air speed had the settings $v = 0.1, 0.4$ and 1.5 m/s corresponding to off (for winter), low (for summer nights), and high speed respectively and (6) that there were three clothing levels with CLO = 0.3, 0.8 and 1.1 for corresponding to summer, ET* default and winter.

Thirdly, SET* thermostat settings were required to define the new 2050 climates of AccuBatch for NatHERS simulation. These SET* comfort band values were found from the 840,000 records of the SET* output (8 combinations of CLO and v; 3 climates; 4 rooms and 9000 temperature and humidity records) for the 2050 Scarce Resource scenario for the Kitchen/Living room and Bedroom. For the SET*90% comfort approach, the search for naturally conditioned temperatures was carried out for both rooms with a wind speed of 1.5 m/s and a CLO of 0.8 to match ET*, whereas for the SET*80%, the CLO had values of 0.3 and 1.1, and both searches selected moderate humidity values, as shown in Table 1.

Table 1 shows all the results of that analysis with:

- The NatHERS 90% acceptability (ET*90%) thermostats for 1990 climate (in row 1),
- The relative humidity of the SET*90% cooling temperature selection (row 2),
- The thermostat settings for SET*90% (row 3) transformed into naturally conditioned temperatures (not the effective values felt by the occupants),
- Parson's SET* comfort criteria of 'slightly warm/slightly cool' acceptability (SET*80%) temperatures in 2050 (row 4), with the Bedroom cooling thermostat value to be determined (TBD),
- The conservative SET*80% thermostat settings that were assumed for this study (row 5) and
- The relative humidity of the SET*80% cooling temperature selection (row 6)

- The thermostat settings for SET*80% (row 7) transformed into naturally conditioned temperatures.

(3)   SET* energy and star rating calculation

To calculate the star ratings in a changing climate, it was necessary to estimate the location of each scenario, since a house in a slightly warmer climate to a heating-dominated Warm Temperate one having a similar humidity will use less energy for the same star rating due to less heating required (DEWHA 2008a, 133). The location was also needed for the correct house area adjustment factor that prevents small house bias and which depends on home area and climate. Trial and error for several towns found that Perth (PE) and Coffs Harbour (CH) presented sufficiently close annual increases in mean monthly temperatures to that of Adelaide under the Scarce Resource and Extreme Climate Change scenarios, respectively, and so these were adopted for this study. Future research could consider the humidity change as well.

Batch jobs of multiple standard AccuRate energy simulations of the house for the Adelaide 1990 and 2050 climates were run using AccuBatch program v2.2.0.0 dated 8th June 2016 (P. Nagle. 2010. "FW: AccuBatch 2.0.0.0 trial." Personal Correspondence, August 11; D. Chen. 2016. RE: AccuBatch – Personal Correspondence, August 6) for the three comfort approaches of standard NatHERS (ET*90%), SET*90% and SET*80%. The comfort approaches used the thermostats as shown in Table 1 and the comfort bands and six SET* details:

- NatHERS ET*90% has a comfort band of 15–25°C, air speeds $v = 0.1$ and 1.5m/s and clothing level (CLO) = 0.8.
- SET*90% has a corresponding 'transformed' comfort band 14.6–27.3°C, air speeds $v = 0.1$ and 1.5m/s and CLO = 0.8.
- SET*80% has a corresponding 'transformed' comfort band 13.7–34.8°C, air speeds $v = 0.1, 0.4$ and 1.5m/s and CLO = 0.3 and 1.1.

Results

Climate change

Table 2 summarizes the projected temperature increases for both scenarios by the HadGEM2-ES climate model using the CSIRO Climate Futures approach (Shiel et al. 2017), from 1995 to 2050 for the:

- Mean of the monthly maximum daily temperature and
- Mean of the monthly minimum daily temperature.

The Adelaide RCP8.5 projection of the mean monthly daily temperature rise of 1.8 K from 1995 to 2050 is in general agreement with other RCP8.5 projections including temperature rises for

Table 2. Mean changes in the monthly maximum and minimum daily temperatures projected by HadGEM2-ES, from 1995 to 2050 for the Adelaide Sub-Cluster.

Scenario	Mean change in the monthly maximum daily temperature (K)	Mean change in the monthly minimum daily temperature (K)
Scarce Resource (RCP4.5)	1.35	1.12
Extreme Climate Change (RCP8.5)	1.85	1.73

Source: Shiel.

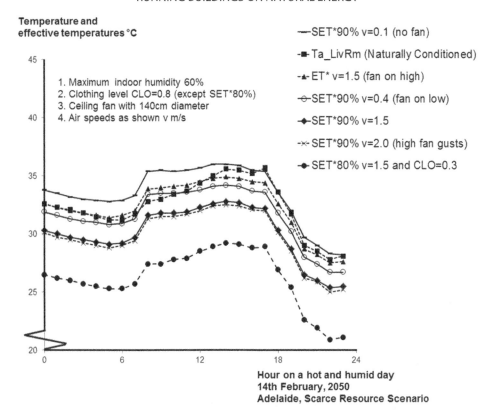

Figure 6. Effective temperatures with different air speeds for a hot and humid day in a 2050 Scarce Resource scenario for Adelaide, showing greater cooling benefit with increasing air speed for the living room of the case study (Source: Shiel).

Sydney's East Coast South sub-cluster of 1.8 K from 1995 to 2050 for the 50th percentile (Dowdy et al. 2015), and for the Australian median surface temperature rise of approximately 1.4 K in a shorter epoch from 2010 to 2050 (CSIRO and BoM 2015, 92).

Effective temperatures, air movement and the cooling effect

Figure 6 shows the benefit of air movement in the Living Room of the case study, for a hot and humid day in the 2050 Scarce Resource scenario for Adelaide. It has the Living Room temperatures and effective occupant temperatures (e.g. with the air speed of SET*90% increasing from $v = 0.1$–2.0 m/s) all with a clothing level of CLO = 0.8, except for SET*80% with $v = 1.5$ and a lower clothing level of CLO = 0.3, with ET* and SET* calculated from the Fountain and Huizenga SET* program.

Simulated energy

Figure 7 shows the case study simulated required energy for comfort and star ratings for three comfort approaches, for the Adelaide climate of 1990 and two climate change scenarios in 2050. It shows for the case study that:

- For the short dashed line at the top (NatHERS) comfort approach, the comfort and star ratings performance of 7 stars deteriorated to 5.9 and then 4.2 stars for the Scarce Resource and Extreme Climate Change scenarios in 2050, respectively.
- For the longer dashed line in the middle (SET*90%) comfort approach, the comfort and star ratings performance starts higher at 9.2 stars but also deteriorates to 8.7 and 7.9 stars for the Scarce Resource and Extreme Climate Change scenarios in 2050, respectively.

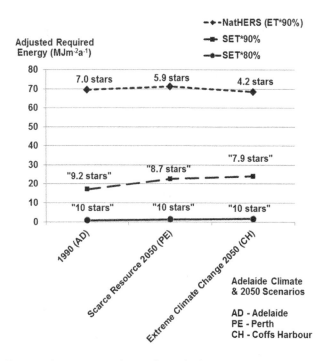

Figure 7. The star ratings and energy for comfort for the case study for three comfort approaches (the NatHERS ET*90%, SET*90% and SET*80%) for three climates: Adelaide 1990, and for two climate change scenarios in 2050 (Scarce Resources with Perth (PE) and Extreme Climate Change with Coffs Harbour (CH)) (Source: Shiel).

- For the full line at the bottom (SET*80%), the NatHERS required energy for comfort is reduced by more than 95% in the 1990 climate, and in the two 2050 climate scenarios. The star rating also remains constant at 10 stars.

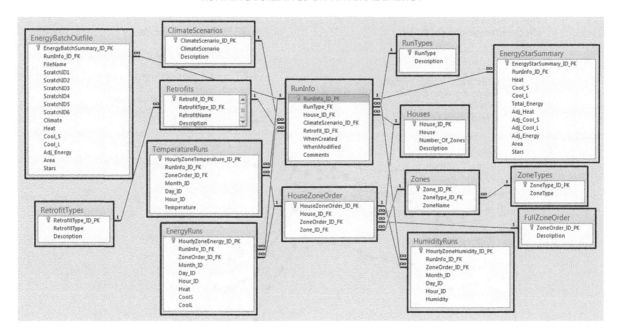

Figure 8. Database design for the simulation results (Source: Shiel).

Database

A results database was created as part of the research to manage hourly temperatures, humidity and energy values for each zone for naturally conditioned and conditioned NatHERS simulations, for different houses and climates. This was accomplished after designing a naming convention for the retrofits, and applying it to AccuRate's scratch and output files, designing and implementing the database with the design shown in Figure 8. A programmer was engaged to implement a database loading program to load all the results files directly from one folder.

Discussion

Case study

Method

The success of the SET* comfort approach comes from its use of all six parameters influencing thermal comfort, and including a model of skin temperature and skin wettedness, the latter of which is directly related to comfort during sweating. The SET* calculation can be performed for occupants, but an acceptable comfort band is also required to know when comfortable or safe health limits have been surpassed. These limits are uncertain and need more research with the absence of long-term acclimatization research (Hanna and Tait 2015, 27) by dwelling field studies with comfort sensation votes, and measurements of physiological heat stress and SET* parameters.

In row 4 of Table 1 for the Bedroom cooling thermostat, the TBD (to be determined) was required because bedroom maximum temperatures required additional research. It was set to 26°C based on Peeters et al. (2009), although this was without a fan, and Cunnington's average of the upper level of the best temperature sleeping range of 24°C and the 28°C that can be tolerated if acclimatized in the Tropics, the maximum sleeping temperature:

> ... people seem to sleep best when bedroom temperatures are between 16–24 degrees Celsius. Those who are used to sleeping in

warmer temperatures such as when living in the tropics can acclimatise to a certain degree and sleep reasonably in bedroom temperatures up to 28 degrees Celsius. (Cunnington 2016)

The selection of the naturally conditioned temperatures that correspond to the SET* 2050 index values needs to be conservative because:

- the transformation is a novel approach and estimates the SET* required energy based on transforming temperatures;
- The temperatures have different definitions (environmental and operative),
- The CHENATH engine has many complexities that may be more suited to the ET*90% approach, for example, the fan on and off temperatures, which were changed in AccuBatch from 24.5°C and 22°C, respectively, to 27°C and 24°C, respectively, for this study for both SET*90% and SET*80%.

Climate change scenarios

Two climate change scenarios were developed for this research programme. The Extreme Climate Change scenario was developed since it is similar to the current temperature trajectory, and assumes large reserves of fossil fuels and expanding energy, population and continued high consumption rates and emissions (IPCC-AR5-WGI 2013, 104).

The Scarce Resource scenario was developed to take into account slower economic growth patterns beginning to be evident today, as the consumption patterns of a burgeoning middle class take effect on the planet's finite resources. It models a world of resource depletion, especially with regard to per capita freshwater, soil, arable land, oil and certain minerals affecting construction and energy materials and prices, leading to lower economic growth, consumption and emissions. (Pfeiffer 2006; Heinberg 2007; Kempf 2008; Victor 2008; Hall 2010; Larsson 2010; Bol 2011; McKinsey & Company 2011; Hall 2012; Klare 2012; Rubin 2012; Turner 2012; Tverberg 2014; Mohr et al. 2015; Steffen et al. 2015; Biswas 2016; WEC 2016).

Effective temperatures, air movement and the cooling effect

Figure 6 shows the case-study Living Room temperatures on a hot and humid day in 2050 where:

- The naturally conditioned temperature reaches 36°C and the relative humidity reaches a maximum of around 60%;
- Greater air movement lowers the effective temperatures of the occupants as the SET*90% air speeds increase from $v = 0.4$–2.0 m/s, although SET*90% $v = 2$ m/s has almost the same benefit as SET*90% $v = 1.5$ m/s.
- The NatHERS ET*90% temperature should be around the ET* $v = 1.5$ and the SET*90% $v = 1.5$ values in Figure 6 (which both do not use the Szokolay cooling effect) and would require conditioning when it rises above the 27.3°C Adelaide cooling thermostat, assuming the 2.8 K maximum Szokolay effect for a 140 cm fan;
- The wider SET*80% comfort temperature band with the CLO = 0.3 (e.g. shorts, shirt and shoes) keeps the temperature that the occupants feel lower than the thermostat trigger of 28°C for most of day with little energy needed for conditioning (see the grey shaded area of Figure 5) compared with the Living Room temperature and the NatHERS thermostat level from 7 am to midnight).

Around 8 am in Figure 6 the effective temperatures for 3 comfort approaches (SET*90% $v = 0.1$, ET* $v = 1.5$ and SET*90% $v = 0.4$) exceed the naturally conditioned temperature Ta_LivRm. This is because the humidity is around 60% and these effective temperatures do not provide as much cooling effect as SET*90% $v = 1.5$ or SET*80% $v = 1.5$, the latter of which includes a lower CLO.

Energy savings

Figure 7 shows that even dwellings well designed for the Adelaide climate of 1990 will be uncomfortable in 2050 using the NatHERS ET*90% comfort approach, and this may encourage more air-conditioning usage.

The SET*90% comfort approach provides better results with the slightly wider comfort temperature band of 14.6–27.3°C compared to 15–25°C for NatHERS since it takes thermoregulation and skin wettedness into account at higher temperatures and humidity, whereas NatHERS uses the cooling approach of Szokolay.

The comfort approach for SET*80% showed exceptional results with varied clothing levels and using an extra fan speed allowing the wide comfort band of 13.7–34.8°C (which feels like 16–28°C for the appropriate air speed, humidity, light and heavy clothing, and other parameters).

Existing stock of dwellings

Figure 7 suggests that deep retrofits would be required for the existing low thermal performance housing stock for 2050 if the Extreme Climate Change scenario eventuated for the NatHERS ET*90% and SET*90% comfort approaches, whereas more urgent retrofits would be needed for the Scarce Resource scenario. However, the results of the lowest two lines in Figure 7 suggest that well-designed homes may not require retrofits if the occupants actively manage the home, but this may require a

programme of education. This suggests that the SET*80% comfort approach could be the most effective one for global warming for adaptation, and maintaining occupant comfort, as well as for mitigation and to reduce carbon emissions.

The case-study ventilation and IEQ

The case study has large ceiling fans in the Bedroom and Living Room/Kitchen if natural ventilation is not available. It was also designed as a tightly sealed house that included sealed exhaust fans in the Kitchen, Laundry and Bathroom to assist with moisture control. If this house was built to the NatHERS assumed level of weather-stripping, the air changes per hour (ACH) would be too low at less than 0.5 ACH at natural pressure (ACHnat) for a healthy environment without an Energy Recovery Ventilation system (Lstiburek 2013). However, occupants can often manage the air exchange rate manually by opening doors and windows as required.

The future of NatHERS comfort

One of the main originators of the AccuRate program, Angelo Delsante, felt that it was more realistic to rate houses on a naturally conditioned basis, rather than cooling requirements, but found that this needed a comfort index and a better ventilation model (Delsante and CSIRO 1997, 11). The Degree Discomfort Hours (DDH) approach has been addressed by various authors (Aynsley and Szokolay 1998; Willrath 1998; Kordjamshidi 2011), but when Aynsley and Szokolay were considering NatHERS comfort approaches in 1998, they found that the SET* index was the most suitable for residential comfort. It was not recommended, however, since few Australians were living in the humid North, SET* algorithms were not readily available, and it required additional computing effort (Aynsley and Szokolay 1998, 21). So ET* was recommended as a subset of SET* and the next most suitable index.

An expert NatHERS workshop group agreed that more existing houses should be monitored and results correlated with AccuRate simulations to validate NatHERS tools, and recommended changes to the cooling thermostats, including bedroom temperatures similar to the one trialled here (Saman et al. 2008).

One recent approach proposed for NatHERS that could improve the cooling effect of air movement (Daniel et al. 2015) extends the ATC approach with a warmer linear indoor operative temperature extension to a region of the outdoor air temperature above 25.3°C. However, this relies on one parameter out of six to calculate comfort and disregards the non-linear evaporative cooling effects of sweating (Chan 2017) and the degree of skin wettedness (Parsons 1993, 167, 212), which limits human thermoregulation and acclimatization capacity (Hanna and Tait 2015, 8038–8044), but is taken into account by SET* (Parsons 1993, 233–237; Aynsley 2008).

SET* could replace ET*, or be included as a dual comfort approach, SET* and ET*, which could work like the dual city and country car fuel economy ratings, and would provide a range of ratings and allow occupants to decide if they are in the 'Active' or 'Passive' category. Including SET* in NatHERS is consistent with Delsante's realistic way to rate houses, with the NatHERS workshop to modify thermostat levels, and the continued support of the two original authors of the NatHERS 1998 comfort approach report (R. Aynsley, Szokolay – Personal Correspondence, January 22, 2017).

Prioritizing retrofits

A national Mandatory Residential Disclosure (MRD) scheme for existing dwellings could help to identify the buildings where retrofits are needed most. This should be subsidized by the government since the results would help:

- To validate and improve NatHERS if both monitoring and simulation were done
- Assist with mitigating and adapting to climate change, where a retrofit programme could be designed after a performance survey of the building stock and include educating occupants about personal comfort controls. The mitigation would assist Australia in meeting its international low carbon targets.

A detailed NatHERS results database

The opportunity was taken to design and develop a NatHERS results database to store the detailed NatHERS hourly zone simulated information, unlike the recently created national NatHERS repository of summary dwelling rating information. It can use SQL queries to retrieve and compare naturally conditioned and rating results such as hourly temperatures, humidities or energy, across houses, climates, zones (e.g. for the Living or Bedroom), hottest and coldest days, summer and winter seasons, and calculate energy savings or DDHs improvements for retrofits over base cases. The database could be extended to include more house attributes including images for plans, photos; monitored climate data and results; Assessor details including education and experience; NABERS ratings and SET temperature values for various clothing and air speed levels for each temperature and humidity to help determine regional human thermoregulatory acclimatization limits.

Some possible database application examples are:

- NatHERS validation by comparing simulated and monitored house results (Saman et al. 2008, 27).
- Colour temperature 'contour' diagrams of a house plan, showing hottest or coldest zones at various times in the year, which could assist in prioritizing building occupant evacuations during heat waves (Roaf, Crichton, and Nicol 2005).
- New design or retrofit advice for dwellings, from similar existing houses stored with good star ratings, for a particular house construction and climate.
- Calculating DDHs with variable thresholds.
- Helping to determine SET* thermostat settings per location.

Conclusions

The conclusions are:

- An estimated saving of 95% of the NatHERS heating and cooling energy for a house case study could be made with one SET* comfort approach, deserving more research for inclusion in NatHERS;

- Adelaide temperatures were projected to increase 1.2 and 1.8 K from 1995 to 2050 for the Scarce Resource and Extreme Climate Change scenarios, respectively;
- Much of the Australian residential stock will require rapid retrofits if a Scarce Resource scenario unfolds and prices rise, or deep retrofits if an Extreme Climate Change scenario eventuates;
- A cost-effective education programme of 'Active' adaptive behaviour for occupants of well-designed dwellings could avoid the need for retrofits to these dwellings;
- A subsidized national MRD scheme should be implemented; and
- A detailed NatHERS simulation results database has been created which could be of value to assessors, other researchers and NatHERS.

Further research areas could include field studies for SET* acclimatization comfort bands incorporating thermoregulation testing of subjects; maximum sleeping temperatures and the database applications identified.

Acknowledgements

The authors would like to thank the CSIRO NatHERS simulation team and especially Dr Dong Chen for his help and discussions; CSIRO's Dr John Clarke for assistance with climate modelling; Dr Dariusz Alterman for an earlier review and Dr Karen Blackmore for database normalization assistance, both from the University of Newcastle; Alan Leslie for his Python programming and IT skills and in particular the reviewers for their insightful comments.

Disclosure statement

No potential conflict of interest was reported by the authors.

ORCID

John J. Shiel ⓘ http://orcid.org/0000-0002-3233-6284

References

ABS. 2013a. *3101.0 – Australian Demographic Statistics*. Australian Bureau of Statistics. http://www.abs.gov.au/ausstats/abs@.nsf/mf/3101.0.

ABS. 2013b. *8752.0 – Building Activity, Australia, Feature Article: Average Floor Area of New Residential Dwellings, June 2013*. 8752.0. Australian Bureau of Statistics. http://www.abs.gov.au/AUSSTATS/abs@.nsf/Previousproducts/8752.0Feature%20Article1Jun%202013?opendocument&tabname=Summary&prodno=8752.0&issue=Jun%202013&num=&view=.

Ambrose, M., M. James, A. Law, P. Osman, S. White, and CoA-DI. 2013. *The Evaluation of the 5-Star Energy Efficiency Standard for Residential Buildings*. CSIRO, Commonwealth of Australia, Department of Industry. http://www.industry.gov.au/Energy/Pages/Evaluation5StarEEfficiencyStandardResidentialBuildings.aspx.

ASBEC. 2008. *The Second Plank – Building a Low Carbon Economy with Energy Efficient Buildings*. Flemington: The Australian Sustainable Built Environment Council (ASBEC). http://www.asbec.asn.au/research.

ASBEC. 2016. *Low Carbon, High Performance – How Buildings Can Make a Major Contribution to Australia's Emissions and Productivity Goals*. Surry Hills: Australian Sustainable Built Environment Council. http://www.asbec.asn.au/research/.

ASHRAE-55. 2010. *ANSI/ASHRAE Standard 55-2010 – Thermal Environmental Conditions for Human Occupancy*. 55. Atlanta, GA: ASHRAE.

Auliciems, A., S. V. Szokolay, UoQ, and PLEA. 1997. *Thermal Comfort*. Design Tools and Techniques; Note 3, PLEA Notes. Brisbane: International Passive and Low Energy Architecture (PLEA) Organisation, and Dept. of Architecture, University of Queensland.

Aynsley, R. 2012. "Condensation in Residential Buildings. Part 1 Review." *Ecolibrium* 11 (9): 48–52.

Aynsley, R. M. 2008. "Quantifying the Cooling Sensation of Air Movement." *International Journal of Ventilation* 7 (1): 67–76.

Aynsley, R. M., and S. V. Szokolay. 1998. *Options for Assessment of Thermal Comfort/Discomfort for Aggregation into NatHERS Star Ratings*. Townsville: James Cook University.

Baharun, A., K. Ooi, and D. Chen. 2009. "Thermal Comfort and Occupant Behaviors in Accurate, A Software Assessing the Thermal Performance of Residential Buildings in Australia." In *Proceedings of the 3rd International Conference on Built Environment and Public Health, EERB-BEPH, Guilin, China*.

Belcher, S., J. Hacker, and D. Powell. 2005. "Constructing Design Weather Data for Future Climates." *Building Services Engineering Research and Technology* 26 (1): 49–61.

Biswas, S. 2016. "Is India Facing Its Worst-ever Water Crisis?" *BBC News*, March 27. http://www.bbc.com/news/world-asia-india-35888535.

Blumer, C., and L. Mayers. 2017. "Sydney Sweats Through Hottest January Night Ever." Text. *ABC News*. http://www.abc.net.au/news/2017-01-14/sydney-heat-breaks-another-hottest-night-record/8182602.

Bol, D. 2011. "Material Scarcity and Its Effects on Energy Solutions." Brussels, Belgium. http://www.aspo.be/assets/ASPO9_Wed_27_April_Aleklett.pdf.

BoM. 2016. "Bureau of Meteorology's Climate Data Online – Weather and Climate Statistics – Monthly Temperatures – Period 1981–2010 – Sydney Observatory Hill; Brisbane Archerfield Airport." http://www.bom.gov.au/climate/data/index.shtml.

Byford, H., A. Hage, and CRC-LCL. 2016. "Launch of RP1024 NextGen Ratings Tool Project. Low Carbon Living CRC." *Launch of RP1024 NextGen Ratings Tool Project*. http://www.lowcarbonlivingcrc.com.au/news/news-archive/2016/09/unisa-media-release-launch-rp1024-nextgen-ratings-tool-project.

BZE. 2013. *Zero Carbon Australia Buildings Plan*. Melbourne: Beyond Zero Emissions. http://bze.org.au/buildings.

Candido, C. 2011. "Adaptive Comfort: Passive Design for Active Occupants." *Environment Design Guide* 69 (Sept.): 1–5.

Chan, A. 2017. "Heat Exchange Between the Human Body and the Environment." *Heat Exchange*. Accessed January 21. http://personal.cityu.edu.hk/bsapplec/heat.htm.

Chen, D. 2008. "The Latest in Software Innovation. Update on AccuRate Development." Presented at the ABSA National Conference, Melbourne, Australia. http://www.hearne.com.au/attachments/AccuRate%20presen-tation%20by%20Dr.%20Dong%20Chen%20CSIRO.pdf.

Chen, D. 2014. "NatHERS Update 2014 – An Introduction to the New Software Engine Release Commencing Use on 1 October 2014 and Details of the New Software Output Certificate." Presented at the NatHERS House Energy Rating Scheme Update.

Chen, D., and CSIRO. 2016. "AccuRate and the Chenath Engine for Residential House Energy Rating." https://www.hstar.com.au/Home/Chenath.

Chen, D., X. Wang, and Z. Ren. 2012. "Selection of Climatic Variables and Time Scales for Future Weather Preparation in Building Heating and Cooling Energy Predictions." *Energy and Buildings* 51: 223–233. doi:10.1016/j.enbuild.2012.05.017.

Copper, J. 2012. "Measurement and Verification of Solar Irradiance and Residential Building Simulation Models for Australian Climates." Sydney: University of New South Wales. http://unsworks.unsw.edu.au/fapi/datastream/unsworks:11065/SOURCE01?view=true.

CSIRO. 2015. "Australian Climate Futures Online Tool – Climate Change in Australia. Projections for Australia's NRM Regions." http://www.climatechangeinaustralia.gov.au/en/climate-projections/climate-futures-tool/introduction-climate-futures/.

CSIRO, and BoM. 2015. *Climate Change in Australia – Projections for Australia's Natural Resource Management Regions: Technical Report*. CSIRO and Bureau of Meteorology. http://www.climatechangeinaustralia.gov.au/en/publications-library/technical-report/.

Cunnington, D. 2016. "Sleeping in the Heat. Section: Why Is It Hard to Sleep in the Heat?." *SleepHub*. http://sleephub.com.au/sleeping-in-the-heat/.

Daniel, L., T. Williamson, V. Soebarto, and D. Chen. 2015. "A Model for the Cooling Effect of Air Movement." In *Proc. 49th Annual Conference of ANZAScA*. Melbourne Victoria: The Architectural Science Association and the University of Melbourne. http://anzasca.net/wp-content/uploads/2015/12/103_Daniel_Williamson_Soebarto_Chen_ASA2015.pdf.

de Dear, R., G. Brager, J. Reardon, and F. Nicol. 1998. *Developing an Adaptive Model of Thermal Comfort and Preference/Discussion*. ProQuest. http://search.proquest.com/openview/bd3427db1cb55e6e9ab20d3099a6d8e4/1?pq-origsite=gscholar.

Delsante, A. 2005. "Is the New Generation of Building Energy Rating Software Up to the Task? A Review of AccuRate." *ABCB Conference Building Australia's Future*.

Delsante, A. E., and CSIRO. 1997. *The Development of an Hourly Thermal Simulation Program for Use in the Australian Nationwide House Energy Rating Scheme*. CSIRO. Division of Building, Construction and Engineering. http://www.inive.org/members_area/medias/pdf/Inive%5Cclima2000%5C1997%5CP354.pdf.

DEWHA. 2008a. *Energy Use in the Australian Residential Sector 1986–2020*. Department of the Environment, Water, Heritage and the Arts. http://industry.gov.au/Energy/EnergyEfficiency/StrategiesInitiatives/NationalConstructionCode/Documents/energyuseaustralianresidentialsector198662020part1.pdf.

DEWHA. 2008b. *Your Home – Design for Lifestyle and the Future – Technical Manual*. 4th ed. Paragon Printers Australasia. http://www.yourhome.gov.au/technical/index.html.

Dewsbury, M. 2011. "The Empirical Validation of House Energy Rating (HER) Software for Lightweight Housing in Cool Temperate Climates." Hobart: University of Tasmania. http://eprints.utas.edu.au/12431/.

Dowdy, A., J. Bhend, F. Chiew, J. Church, J. Ekstrom, D. Kirono, A. Lenton, et al. 2015. *East Coast Cluster Report. Regions: Cluster Reports. Climate Change in Australia Projections for Australia's Natural Resource Management*. CSIRO & Bureau of Meteorology. https://www.climatechangeinaustralia.gov.au/.

Fanger, P. O. 1972. *Thermal Comfort: Analysis and Applications in Environmental Engineering*. New York: McGraw-Hill.

Fountain, M., and C. Huizenga. 1995. *A Thermal Sensation Prediction Model for Use by the Engineering Profession*. ASHRAE 781RP. Peidmont, CA: ASHRAE. http://escholarship.org/uc/item/3g98q2vw.

Gagge, A. P., A. P. Fobelets, and L. G. Berglund. 1986. "A Standard Predictive Index of Human Response to the Thermal Environment." *ASHRAE Transactions* 92 (2B): 709–731.

Grimme, F., M. Laar, and C. Moore. 2003. "Man & Climate – Are We Losing Our Climate Adaptation?" In *Proc. RIO3 – World Climate & Energy, Rio de Janeiro, Brazil*.

Hacker, J., J. De Saulles, A. Minson, and M. Holmes. 2008. "Embodied and Operational Carbon Dioxide Emissions from Housing: A Case Study on the

Effects of Thermal Mass and Climate Change." *Energy and Buildings* 40 (3): 375–384.

Hall, C. A. S. 2012. "Energy Return on Energy Invested." In *The Energy Reader: Overdevelopment and the Delusion of Endless Growth*. Healdsburg, CA: Watershed Media. http://energy-reality.org/wp-content/uploads/2013/05/09_Energy-Return-on-Investment_R1_012913.pdf.

Hall, M. R. 2010. *Materials for Energy Efficiency and Thermal Comfort in Buildings*. 1st ed. *Woodhead Publishing Series in Energy 14*. Boca Raton, FL: CRC Press.

Hanna, E. G., and P. W. Tait. 2015. "Limitations to Thermoregulation and Acclimatization Challenge Human Adaptation to Global Warming." *International Journal of Environmental Research and Public Health* 12 (7): 8034–8074.

Heinberg, R. 2007. *Peak Everything: Waking Up to the Century of Declines. FEP Torn*. Gabriola, BC: New Society Publishers.

Henriksen, J. 2005. "The Value of Design in Reducing Energy Use and CO2-E Impact over the Life Cycle of a Detached Dwelling in a Temperate Climate." University of Newcastle.

Humphreys, M., F. Nicol, and S. Roaf. 2015. *Adaptive Thermal Comfort: Foundations and Analysis*. London: Routledge.

IPCC-AR5-WGI. 2013. *Climate Change 2013: The Physical Science Basis. Contribution of Working Group I to the Fifth Assessment Report (AR5) of the Intergovernmental Panel on Climate Change*, edited by T. F. Stocker, D. Qin, G. K. Plattner, M. Tignor, S. K. Allen, J. Boschung, A. Nauels, Y. Xia, V. Bex, and P. M. Midgley. New York: Cambridge University Press. http://www.ipcc.ch/report/ar5/wg1/.

Janda, K. B. 2011. "Buildings Don't Use Energy: People Do." *Architectural Science Review* 54 (1): 15–22. doi:10.3763/asre.2009.0050.

Kempf, H. 2008. *How the Rich Are Destroying the Earth*. White River Junction, Vt: Green Books.

Klare, M. 2012. *The Race for What's Left: The Global Scramble for the World's Last Resources*. Reprint, New York: Picador.

Kordjamshidi, M. 2011. *House Rating Schemes from Energy to Comfort Base*. Berlin: Springer.

Larsson, N. 2010. "Rapid GHG Reductions in the Built Environment under Extreme Conditions." *International Journal of Sustainable Building Technology and Urban Development* 1 (1): 15–21. doi:10.5390/SUSB.2010.1.1.015.

Lstiburek, J. 2013. "Deal with the Manure and Then Don't Suck." *ASHRAE Journal*, July. http://bookstore.ashrae.biz/journal/download.php?file = 2013 July-040-046_building-sciences_lstiburek.pdf.

McKinsey & Company. 2011. *Resource Revolution: Meeting the World's Energy, Materials, Food, and Water Needs*. Seoul: McKinsey Global Institute. http://www.mckinsey.com/insights/mgi/research/natural_resources/resource_revolution.

Mohr, S. H., J. Wang, G. Ellem, J. Ward, and D. Giurco. 2015. "Projection of World Fossil Fuels by Country." *Fuel* 141 (February): 120–135. doi:10.1016/j.fuel.2014.10.030.

Moore, T., I. Ridley, Y. Strengers, C. Maller, and R. Horne. 2016. "Dwelling Performance and Adaptive Summer Comfort in Low-income Australian Households." *Building Research & Information* 1–14. doi:10.1080/0961 3218.2016.1139906.

Nicol, F., M. Humphreys, and S. Roaf. 2012. *Adaptive Thermal Comfort: Principles and Practice*. 1st ed. London: Routledge.

Page, A. W., B. Moghtaderi, D. Alterman, and S. Hands. 2011. *A Study of the Thermal Performance of Australian Housing*. Callaghan: Priority Research Centre for Energy, University of Newcastle. http://www.thinkbrick.com.au/download.php?link = assets/Uploads/TB-PHASEI-REPORT-FINAL-web.pdf&name = TB-PHASEI-REPORT-FINAL-web.pdf.

Parsons, K. C. 1993. *Human Thermal Environments: The Effects of Hot, Moderate, and Cold Environments on Human Health, Comfort, and Performance*. 2nd ed. Boca Raton: CRC Press.

Peeters, L., R. de Dear, J. Hensen, and W. D'haeseleer. 2009. "Thermal Comfort in Residential Buildings: Comfort Values and Scales for Building Energy Simulation." *Applied Energy* 86 (5): 772–780. doi:10.1016/j.apenergy.2008.07.011.

Pfeiffer, D. 2006. *Eating Fossil Fuels: Oil, Food and the Coming Crisis in Agriculture*. Gabriola Island: New Society Publishers.

Ren, Z., and Z. Chen. 2010. "Enhanced Air Flow Modelling for AccuRate – A Nationwide House Energy Rating Tool in Australia." *Building and Environment* 45 (5): 1276–1286. doi:10.1016/j.buildenv.2009.11.007.

Ren, Z., and D. Chen. 2015. "Estimation of Air Infiltration for Australian Housing Energy Analysis." *Journal of Building Physics* 69–96. doi:10.1177/1744259114554970.

Ren, Z., Z. Chen, and X. Wang. 2011. "Climate Change Adaptation Pathways for Australian Residential Buildings." *Building and Environment* 46 (11): 2398–2412. doi:10.1016/j.buildenv.2011.05.022.

Ren, Z., X. Wang, and D. Chen. 2014. "Heat Stress Within Energy Efficient Dwellings in Australia." *Architectural Science Review* 57 (3): 227–236. doi:10.1080/00038628.2014.903568.

Roaf, S., D. Crichton, and F. Nicol. 2005. *Adapting Buildings and Cities for Climate Change: A 21st Century Survival Guide*. Amsterdam: Architectural Press.

Romm, C. 2016. "Too Much Time in Air-conditioning Is Warping Your Ability to Handle Heat." *New York Magazine – Science of Us*, December 8. http://nymag.com/scienceofus/2016/08/too-much-air-conditioning-is-warping-how-you-handle-heat.html.

Rubin, J. 2012. *The End of Growth*. Toronto: Random House Canada.

Ryan, P., and M. Pavia. 2016. "Australian Residential Energy End-use: Trends and Projections to 2030." In *Proc. ACEEE Summer Study on Energy Efficiency in Buildings*. Pacific Grove, CA: American Council for an Energy-Efficient Economy (ACEEE). http://aceee.org/files/proceedings/2016/data/papers/9_286.pdf#page = 1.

Saman, W., NCCARF, and UniSA. 2013. *A Framework for Adaptation of Australian Households to Heat Waves*. Gold Coast: National Climate Change Adaptation Research Facility (Australia). http://hdl.handle.net/10462/pdf/3250.

Saman, W., M. Oliphant, L. Mudge, and E. Halawa. 2008. *Study of the Effect of Temperature Settings on AccuRate Cooling Energy Requirements and Comparison with Monitored Data*. Residential Building Sustainability, DEWHA. https://www.hearne.com.au/getattachment/e7fd4375-7cb2-4ea9-991e-a6de43fb2641/AccuRate%20predictions%20compared%20with%20monitored%20data.aspx.

Shiel, J., B. Moghtaderi, R. Aynsley, and A. Page. 2014. "Reducing the Energy Consumption of Existing Residential Buildings for Climate Change and Scarce Resource Scenarios in 2050." In *Weather Matters for Energy*. New York: Springer. http://www.springer.com/us/book/978146149 2207.

Shiel, J., B. Moghtaderi, R. Aynsley, A. Page, and J. Clarke. 2017. "Rapid Decarbonisation of Australian Housing in Warm Temperate Climatic Regions for 2050." In *Proc. of World Renewable Energy Congress XVI, Perth, WA, Australia*. http://www.wrec2017.com/.

Steffen, W., W. Broadgate, L. Deutsch, O. Gaffney, and C. Ludwig. 2015. "The Trajectory of the Anthropocene: The Great Acceleration." *The Anthropocene Review* 2 (1): 81–98. doi:10.1177/2053019614564785.

Szokolay, S. V. 2004. *Introduction to Architectural Science: The Basis of Sustainable Design*. Amsterdam: Elsevier, Architectural Press.

Turner, G. M. 2012. "On the Cusp of Global Collapse? Updated Comparison of the Limits to Growth with Historical Data." *GAIA: Ecological Perspectives for Science & Society* 21 (2): 116–124.

Tverberg, G. 2014. "Limits to Growth – At Our Doorstep, But Not Recognized." *Our Finite World*. http://ourfiniteworld.com/2014/02/06/limits-to-growth-at-our-doorstep-but-not-recognized/.

Victor, P. A. 2008. *Managing Without Growth: Slower by Design, Not Disaster*. Cheltenham: Edward Elgar Publishing.

Walsh, P. J., and A. Delsante. 1983. "Calculation of the Thermal Behaviour of Multi-Zone Buildings." *Energy and Buildings* 5: 231–242.

WEC. 2016. *World Energy Trilemma 2016: Defining Measures to Accelerate the Energy Transition*. London: World Energy Council. https://www.worldenergy.org/publications/2016/world-energy-trilemma-2016-defining-measures-to-accelerate-the-energy-transition/.

White, D. 2009. "'Passive' Building Design and Active Inhabitants: The Potential of Frugal Hedonism?" In *26th Conf. on Passive and Low Energy Architecture*. Quebec, Canada: Université Laval, Quebec. http://www.plea2009.arc.ulaval.ca/En/Proceedings.html.

Williamson, T., V. Soebarto, and A. Radford. 2010. "Comfort and Energy Use in Five Australian Award-winning Houses: Regulated, Measured and Perceived." *Building Research & Information* 38 (5): 509–529.

Willrath, H. 1998. "The Thermal Performance of Houses in Australian Climates." University of Queensland.

Saving energy with a better indoor environment

Gary J. Raw, Clare Littleford and Liz Clery

ABSTRACT

The analysis in this paper is based on literature reviews, qualitative research and a quantitative survey of 2,313 British households. The aim is to enhance understanding of the needs that people meet in using energy. In particular, we seek to connect energy use with indoor environmental quality and comfort in a way that can promote both energy savings and good indoor environments. We also examine comfort not simply as the thermal state of individuals but in terms of the actions of groups of individuals, in the context of conflicting needs. We categorize the needs of building users under headings of *Other people*, *Comfort* (not only thermal comfort), *Hygiene*, *Resource* and *Ease*. These factors discriminate among population groups in a way that can be used to inform the design and implementation of heating arrangements for homes.

Introduction

The potential impact of energy-saving activity on quality of life has been identified as a barrier to saving energy where householders perceive that the required changes in energy consumption will entail greater discomfort or some other sacrifice of standard of living (Gatersleben 2001; Lorenzoni, Nicholson-Cole, and Whitmarsh 2007). And yet the indoor environment and energy use do not have to be in conflict. This paper seeks to show that, by understanding all the relevant needs of building users, there is instead the potential to engage users to reduce energy use in ways that meet their needs – including the need for a good quality indoor environment. The context of the research is energy use at home in Great Britain but the principles discussed in the paper apply to other national contexts and other uses of buildings.

The starting point is to recognize different perspectives on energy efficiency. Energy *conservation* entails simply using less energy (i.e. less energy input), whatever the consequences for 'output' (the outcomes of using energy). In contrast, energy *efficiency* should be seen as the ratio of output to input. But how should the input and output be defined and what are the system boundaries? At the level of a single gas boiler, for example, there is input of chemical energy and output of heat energy. From the perspective of boiler design, this is important but the boundaries of the system can be set much wider to include, for example, fuel sourcing, extraction and distribution, hot water distribution within the home, heat emission and heat retention. Efficiency can – and should – be understood in terms of input and output at any and all of these levels.

However, this is all from the perspective of physics and chemistry, which is not necessarily how households perceive efficiency. People may not think explicitly in terms of efficiency but we can conceptualize their thinking in this way in order to show how it differs from an engineering perspective. Households might see inputs as energy or fuel, or more specifically non-renewable or secure energy resources. Alternatively, they might see input as their money, time and space; or physical or mental effort. Output might be seen in relation to, for example, the motives categorized by Raw, Varnham, and Schooling (2010) as relating to the environment ('save the planet', 'save the country' and 'save my household'), the household's resources ('save or make money' and 'avoid waste'), 'well-being', 'improve aesthetics', social motives ('feel good about yourself') and 'make my life easier'. This variation in perspective is important. For example, from a householder's perspective, it may be perceived as efficient to have the heating constantly operating – not because this saves energy but because it saves time and effort while ensuring that the home is always warm enough.

So the scene is complex. It can be ambiguous whether some parameters are input or output, the parameters are difficult to quantify and there are a lot of them. Relationships between physical and psychological quantities are not necessarily linear or even continuous or monotonic. Decisions are not necessarily conscious or – to an observer with a different understanding of efficiency – rational. Households' treatment of input and output parameters is also varied – between persons, between physical, social and cultural contexts, and over time. Nevertheless, an effective intervention to increase household energy efficiency must understand efficiency from the perspective of the household. The intervention has to take motive into account as part a critical trio of *Means*, *Motive* and *Opportunity*.

- *Means* is the technology and/or behaviour that will reduce energy use.
- *Motive* is the reason why households will want to act.
- *Opportunity* is the resource (e.g. time, space or money) to act.

In other words, householders' response to an intervention depends on knowing what to do, having a reason for doing it and having the resources to do it (Raw, Varnham, and Schooling 2010). The biggest complication is to understand and quantify consumers' motives (or the needs they meet, or seek to meet, in using energy). Where a need results in a behaviour, that need can be considered a motive for the behaviour; hence, the two terms are almost interchangeable. Research and practice typically refer to objectively measured energy use but relatively poorly defined and quantified benefits of the energy use (from a user's perspective). But if we understand the motives and needs, the way is open to encourage people to conserve energy in a way that they perceive to be efficient.

The review by Raw and Ross (2011) begins to describe and categorize the motives that underpin use of energy at home. Household resources, well-being and social factors tend to dominate, with wider environmental implications having a supporting role. The details of priority can vary widely between individuals, contexts and the specific behaviours or choices in question, and over time. One motive may result in a range of behaviours and the same behaviours can result from different motives. A need that is currently being met might not be thought of as a need, and the need that is foremost in someone's mind can obscure evidence of other needs. Each motive, individually and in detail, with identified linkages and priorities, should be considered in understanding behaviour and encouraging energy savings.

One key consideration, for example, is the thermal comfort of the occupants of buildings. However, thermal comfort should not be considered in isolation from other needs. In the extreme, research has entailed detailed laboratory analysis of non-interacting individuals (aggregated in analysis and in predictions to represent the comfort of non-interacting groups), studying the impact of hygrothermal parameters, clothing insulation and metabolic rate, with a single subjective comfort rating as the dependent variable. Even field studies exploring the adaptive approach (Nicol, Humphreys, and Roaf 2012) have often seen comfort as a single outcome for individual persons; while the influence of other needs is acknowledged, it is rarely systematically investigated. This paper explores comfort not simply as the thermal state of individuals but in terms of the actions of groups of individuals, in the context of conflicting needs. Our assumption is that what people do in relation to heating the home is related to a set of needs that underlie their use of the home – not just their use of heating. The research therefore begins from the perspective of the needs that drive such energy use and the household-level dynamics that determine how needs are prioritized and actions taken.

This distinguishes the work from much other research that aims to inform the design of heating-related technology and practice – research that takes the physical characteristics of the property as its starting point. Of course, the characteristics of a property (e.g. size, heating system and insulation) set the boundaries of what is possible for people to do when trying to heat their home and keep warm. However, it is essential to understand not only what constrains and enables households' heat energy behaviour, but the underlying goals and motivations that drive and structure that behaviour – including both conscious decisions and the routines and habits that form much of domestic heating behaviour.

To design sustainable heating solutions (either new technologies or different business models), it will be crucial to understand the basic and more complex human needs for heat energy, not just how people currently interact with it. Without this understanding of consumer requirements, future heating solutions could be technically sound but not meet the needs of households. District heating, for example, can provide sustainable, reliable and affordable energy but can be disliked if, for example, it locks the consumer into a single supplier (Upham and Jones 2011) or restricts the household's control over heating (Morgenstern 2012).

The research adopts a broad definition of needs to capture the diversity of people's goals when using energy: anything that people are aiming to achieve through, or what they achieve as a consequence of, using energy. This definition encompasses a wide range of needs, from those objectively essential for life, to preferences based on individual perceived requirements or values. The analysis presented here has started to refine and quantify the categorization of needs suggested by Raw, Varnham, and Schooling (2010) by seeking to understand the different types and roles of needs, and how individual needs may be characterized in terms of underlying dimensions that have the potential to inform the design and implementation of home heating.

Method

Survey sample

The study comprised a quantitative social survey of 2,313 households, which took place in January–February 2014, using quota sampling to generate a nationally representative sample of British households. Quota sampling involves issuing interviewers with a set of quota characteristics and a corresponding number of interviews to be achieved in each category of each characteristic. Its aim is to achieve a representative sample by reflecting the demographic make-up of the areas where interviews are sought. Because quota sampling does not use only random sampling, it is not possible to determine the exact representativeness of a sample.

A total of 250 sample points (COAs – Census Output Areas – containing an average of 300 addresses and derived from the 2011 Census) were randomly selected, covering England, Scotland and Wales. COAs were stratified by Government Office Region (GOR) and a household-level socio-economic indicator – the social grade of household reference person (% Grade A or B – higher or intermediate managerial, administrative or professional role). Sampling units were selected proportionate to the numbers of households within each GOR.

Interviewers given assignments of 10 interviews with quotas in three binary categories: owner vs. renter, house/bungalow vs. flat/maisonette and the presence of children (aged under 18) vs. no children. These quotas were selected because a major literature review (extending earlier reviews, principally Raw and Ross [2011]) and qualitative research (reported in summary by Lipson [2015]) showed them to be closely linked to heat energy needs and behaviours. The quotas numerically reflected sampling unit characteristics (for instance, if, in a given sampling unit, 40% of addresses are owner-occupied, we issued a tenure-based quota of four owner-occupied and six rented addresses). The achieved sample (see Table 1) closely represented the population on the

Table 1. Quota characteristics.

Quota criterion	Quota group	Achieved/expected[a] interviews (%)
Tenure	Owns home	65/66
	Rents home	35/34
Dwelling type	House/bungalow	79/80
	Flat/maisonette	21/20
Presence of children (under 18)	Yes	32/28
	No	68/72

[a]Based on English Housing Survey Headline Report 2011–2012. https://www.gov.uk/government/uploads/system/uploads/attachment_data/file/211288/EHS_Headline_Report_2011-2012.pdf

three issued quotas and other characteristics (such as age of property and size of household).

The sample was based upon household (rather than individual) characteristics because the primary aim was to collect data from individual respondents but relating to their household. For this reason, we aimed to achieve a sample that was diverse with regard to knowledge of heat energy in the household. To this end, there was no systematic respondent selection within households with regard to, for example, who pays the energy bills, who makes more of the energy decisions or who spends most time at home. Interviewers were asked to interview anyone aged over 18 living at the address without selecting (or encouraging self-selection) on the basis of how householders saw their role in relation to energy (although this role was recorded during the interview).

Data collection

The questionnaire design process was iterative and involved extensive collaboration across the research consortium. The literature review and the qualitative research provided considerable learning about past research instruments (e.g. from Raw and Ross [2011]; Shipworth et al. [2010]) and the terminology used by households to discuss heat energy needs and behaviours, which informed the design of survey questions. The full questionnaire was field piloted and certain elements (including identification of heat energy needs) were subject to cognitive testing and redesign.

Respondents in the survey completed a face-to-face computer-assisted personal interview that lasted around 60 minutes, in which they were asked questions relating to household and dwelling demographics, facilities and household behaviour in relation to heating, cooling and hot water, and paying for heat energy. Where respondents consented (89% of cases), interviewers conducted observations of heating and hot water systems. Respondents also received a paper self-completion questionnaire (covering mainly their recent and desired renovation activities); this was returned by 78% of respondents.

Respondents were also asked to complete either two or three card-sort exercises in which they organized a range of pre-defined heat energy needs according to the degree of influence on their heat energy behaviour. The research team defined these needs according to the main categories identified in the literature review and further refined in the pilot survey and cognitive testing. The first card-sort related to (space) heating and keeping warm. This was followed by similar exercises relating to heating water and using hot water, and – if mechanical cooling or ventilation was in use – cooling the home and keeping cool. This paper focuses on the card-sort relating to heating and keeping warm; the 21 card headings and examples are shown in Table 2.

Table 2. Prevalence (%) of needs (*Big Factors* for heating the home and keeping warm).

Heading	Examples given on the cards	%
Being comfortable		85
Energy costs	Not spending more than is necessary.	76
	Keeping the cost of heating under control.	76
Avoiding wasting energy	Not leaving the heating on when it is not needed.	70
Being able to rest and relax		69
Wanting to feel clean	Having a warm room where people can wash and dry themselves.	67
	Having a warm place or radiator to dry laundry.	67
Keeping healthy	Using heat to sooth aches and pains.	61
	Keeping warm to avoid or treat health problems.	61
Feeling in control	Knowing the heating will come on when you want, at the temperature you want.	60
Caring for other members of the household	Making sure the home is warm enough for people (adults or children) with particular needs.	53
Wanting to keep the home clean	Using the heating to avoid damp/mould.	51
	Not using open fires that leave ash or soot.	51
Keeping the home looking, feeling or smelling nice	Avoiding feeling dry or having mould or ugly equipment.	47
	Using fires or heaters to make the home appear cosy.	47
Wanting to feel safe and secure	Not using heating that you worry might be unsafe.	47
	Switching heating systems off when no-one is at home because of safety concerns.	47
The value or cost of your home	Preventing damage to your property that might cost you money.	41
	Installing heating that could increase the value of the home.	41
Concern for the environment	Concern about air pollution, climate change, or the effect of heating on the country's energy resources.	34
Doing what is easiest	Letting the heating controls do the work.	34
Wanting to be productive	Being warm enough to do work at home.	33
The needs of visitors	Ensuring the home is warm enough for visitors.	33
Keeping to your everyday routines	Always having the heating come on at the same time.	28
Doing what you have traditionally done	Doing what you did in previous homes.	16
How you and your home appear to other people	How the temperature of your home appears to other people.	13
	Avoiding appearing either mean or extravagant in your use of heating.	13
Wanting to avoid arguments/disagreements within the home	Avoiding arguments about how warm it is or when the heating is on.	13
Doing what you think most people do	Heating your home in the way you think most people with similar homes would do.	8

The shuffled cards were given to the respondent, who was asked to sort them into three piles according to whether each was a *Big Factor*, *Smaller Factor* or *Not a Factor* in relation to heating and keeping warm. These factors were explained to respondents thus: Big Factors are *very important in influencing what you do*; Smaller Factors are *less important but still influence what you do to some extent*; 'Not a Factor' means *something that does not influence what you do or something that is not relevant to you or your current situation*.

We wanted households to consider all their relevant behaviours when they were asked about which needs they were trying to meet. For this reason, the three card-sorts were all conducted after first asking detailed questions about behaviours in the three domains (relating to heating, cooling and hot water). The intention was that their full range of behaviours should be 'top of mind' while sorting needs, increasing the likelihood of respondents taking all relevant issues into account.

There are potential risks to keeping the questionnaire sequence the same for all respondents, in that the first card-sort may influence respondents' choices in subsequent sorts. However, the focus of our analysis here is on the card-sort relating to heating the home, which was completed first. Had we randomized the order of the sorts, the effects of this on the heating the home exercise would have been unpredictable, weakening the analysis of this set of data. Placing all three card-sorts in quick succession, after the other questions about the three domains of energy use, eliminates risk of the card-sorts progressively influencing responses to other questions; it is also quicker and easier for respondents and interviewers and makes it clearer to respondents how the three sorts differ from each other.

A further set of questions addressed household dynamics. Heating is not necessarily decided by a single individual – it could depend to a greater or lesser extent on any or all members of the household, depending (for example) on the overall household dynamics and who is at home at any given time. The qualitative research suggested the existence of three types of household dynamics:

- 'YOU' – heating is operated according to the needs of one or more individuals, such as a young child or elderly person;
- 'ME' – individuals decide independently to operate the heating according to their own needs;
- 'US' – there is some degree of consensus or cooperation around heating decisions.

In the interviews, a series of questions sought to assign households to one of these types (single-person households were automatically placed in a fourth 'JUST ME' group). The first question asked how the household (if more than one person) decides about heating. Four of the answer options were classified directly, as follows:

- *It's largely down to one person* (ME).
- *It's mainly to care for someone who needs to keep warm or cool, for example because of a health condition or age* (YOU).
- *It varies – depending on whose needs are greatest at the time/It depends more-or-less equally on the needs of everyone who is at home at the time* (US).

Any respondent who instead answered either *It depends on the needs of the person deciding at the time how to heat the home* or *Everyone has a say but one or more people's needs have a greater influence than others* was asked how much influence s/he personally has over decisions about heating. Responses were classified as follows:

- *I tend to have the most influence/Someone else tends to have the most influence/Nobody – we all make decisions about it separately* (ME).
- *It varies – different people influence decisions about the heating at different times/We decide together* (US).

Data preparation and analysis

Because the achieved sample closely matched the population on a range of characteristics, the findings presented here are based on unweighted data. 'Don't know' and 'Refusal' responses are included in bases because they can be relevant responses when measuring behaviour, perceptions and needs (e.g. to identify respondents who are unclear about what their needs or usual behaviour are). Some secondary variables were derived by aggregating levels of a variable (e.g. to assign to income quartiles) and/or by combining more than one variable. These derived variables were defined by the examination of frequency distributions (e.g. to merge small groups) and the logic of combining particular groups.

Data analysis used the statistical package SPSS. In addition to descriptive statistics, the multivariate technique, principal components analysis (PCA) was used to identify whether the 21 needs related to heating the home and keeping warm could be reduced to a smaller number of dimensions. PCA is designed to identify, where they exist, underlying unobservable dimensions based upon the associations between variables. While interpretation of results is partly subjective, and cannot identify causality, it can be a powerful way of identifying fundamental drivers of behaviour.

The PCA used a set of 21 binary variables: whether or not a respondent categorized each need as a *Big Factor* in the card-sort for heating the home and keeping warm. A correlation matrix of the 21 variables was examined to check for (a) sufficient correlation that the existence of a smaller number of underlying dimensions incorporating all variables would be a valid possibility and (b) multi-co-linearity (two or more variables being very highly or entirely correlated) which would suggest that they were measuring the same variable, invalidating the assumptions of PCA. The variables were found to be suitable for PCA.

Results

Heat energy needs

The card-sort was completed by 2287 respondents, who identified a mean of 9.4 *Big Factors* (see Table 2 for details), a mean of 5.5 *Smaller Factors* and a mean of 5.3 *Not a Factor*. Therefore, British households report that they are trying to meet a large number of needs; this needs to be understood when proposing alternative ways of heating the home and keeping warm. Further analysis focuses on the binary variable of whether or not a need is categorized as a *Big Factor*; this is because:

- having more than two groups in further analysis would sometimes require arbitrary assignment of numerical values to *Big Factors*, *Smaller Factors* and *Not a Factor*;
- the number of *Big Factors* typically selected suggests that the *Smaller Factors* were genuinely small;
- the frequency distribution of *Big Factors* is close to normal and indicates that this measure is closest to providing an even split of the sample;
- follow-up questions, seeking to divide the *Big Factors* into two groups, reduced the potential for including all needs in further analysis.

Using *Big Factors* as the principal outcome allows further analysis to account for diversity in numbers of heat energy needs reported by different households. It also ensures that any underlying dimensions identified do not over-simplify a complex picture. All the needs were identified as *Big Factors* by a proportion of respondents. If far fewer needs had been identified as *Big Factors*, it might have been necessary to include needs identified as *Smaller Factors*. However, the statistics indicate that the *Big Factors* are sufficient for the task and we therefore concentrate on these.

The number of needs categorized as *Big Factors* ranged from zero (1.4% of households) to 21 (0.2%) with 6% or more selecting each number of needs from 5 to 12. The mode was 10 needs, selected by 10.3% of respondents. The prevalence of the particular 21 heat energy needs selected as *Big Factors* varies substantially – see Table 2.

More than two-thirds of households identified five particular heat energy needs as *Big Factors*: being comfortable, energy costs, avoiding wasting energy, being able to rest and relax, and wanting to feel clean. This supports the findings of the qualitative research, which argued that comfort and energy costs were primary needs when heating the home for all types of households and that other needs were not considered substantially until these needs had been met. The qualitative research also identified health as a primary need; the survey findings confirm that it is important although in the sixth place behind the needs mentioned above (most likely because it is a met need in most cases).

Some of the 21 needs are relatively rarely prioritized – particularly those classified in qualitative work (Lipson 2015) as sitting in categories of need defined as 'Agency' (the capacity of a person to act independently, and make choices) or 'Relational dynamics'. Less than 30% of households indicated that *Big Factors*, when deciding how to heat the home, included keeping to everyday routines, doing what has traditionally been done, how they and their home appeared to other people, wanting to avoid arguments within the home or doing what they thought most people do. This may, however, reflect a general tendency for people to believe that they are not influenced by what others think or do; in the qualitative research, these needs were not necessarily 'top of mind', but arose from in-depth discussions with respondents. The qualitative research also characterized these needs as more peripheral than the 'core' needs that were more frequently identified as *Big Factors* in the survey.

Dimensions of need

The PCA suggested that five underlying factors (dimensions) of need exist for heating the home; in combination, these dimensions explain 44% of the variance in the data. This indicates that more than half of the variance in the data cannot be explained by the five underlying dimensions and, in effect, does not fit into a neat pattern or a series of patterns across the population as a whole. This reflects both random variance and the sheer diversity of the range and balance of heat energy needs across different households. Nevertheless, such levels of unexplained variance are fairly common for models of this type and a review of relevant statistics suggested that this factor solution gave a very effective summary of the underlying variables. The Kaiser–Meyer–Olkin statistic (a measure of sampling adequacy) is 0.83 – with the literature defining between 0.7 and 0.8 as 'good' and any higher figure as very good. Bartlett's test of sphericity – which tests whether there is some relationship between the variables we want to include in the analysis – produced a significance level of < 0.001, indicating that we can be confident that this is the case. Each factor has an eigenvalue of at least 1.1 whereas subsequent possible factors had eigenvalues of less than 1.0. The five dimensions and the individual heat energy needs that they encapsulate are presented in Table 3.

The range of needs that contribute to each of the five dimensions suggested that these dimensions can be labelled as *Other people*, *Comfort*, *Hygiene*, *Resource* and *Ease* (these labels aim to capture the essence of the dimension but they are more useful as a shorthand to refer to the dimensions rather than as a definitive description). These five 'OCHRE' dimensions account for between 7.4% and 11.0% of the variance and they can be characterized as follows:

- *Other people* is defined by five needs: caring for other members of the household, wanting to avoid arguments within home, the needs of visitors, how you and your home appear to other people, and wanting to be productive. This dimension represents a concern for other people, within or outside the household. The inclusion of productivity is intriguing, suggesting that this need may be interpreted in relation to some mix of supporting others through work in the home and being able to get on with such work, facilitated by cordial relationships and mutual support.
- *Comfort* is defined by three needs: being comfortable, feeling in control, and being able to rest and relax. Whereas the specific need, 'being comfortable' may relate primarily to thermal comfort, the *Comfort* dimension has broader connotations. Being in control (e.g. of the heating system or how other household members use it) is relevant to achieving a certain level of physical comfort. Feeling in control is more than this because it can be difficult to feel comfortable (in the widest sense) in a world (or thermal environment) that is out of control.
- *Hygiene* is defined by five needs: wanting to feel clean, wanting to keep the home clean, keeping the home looking/feeling/smelling nice, keeping healthy, and wanting to feel safe and secure. It represents hygiene in both the specific modern sense of cleanliness and the broader (original) sense of healthiness. It also relates to Herzberg's two-factor

Table 3. Underlying dimensions in relation to heating the home and keeping warm, with their links to individual needs

Specific needs	Dimension of need				
	Other people	Comfort	Hygiene	Resource	Ease
Caring for other members of household	+++				
Wanting to be productive	+				
The needs of visitors	+++				
How you and your home appear to other people	+				
Wanting to avoid arguments/disagreements within the home	++				
Being comfortable		+++			
Feeling in control		+			
Being able to rest and relax		+++			
Wanting to feel clean			++		
Keeping healthy			+		
Wanting to keep the home clean			++		
Wanting to feel safe and secure			+		
Keeping the home looking, feeling or smelling nice			++		
Energy costs				+++	
Avoiding wasting energy				+++	
Concern for environment				++	
The value or cost of your home				+	
Doing what is easiest					++
Keeping to everyday routines					++
Doing what have traditionally done					++
Doing what you think most people do					++

Key: + denotes a positive relationship between the specific heat energy need and the underlying dimension of need, interpreted by a component score greater than ±0.2 (++ denotes a score greater than ±0.3 and +++ greater than ±0.4).

theory of occupational psychology, in which 'hygiene factors' (including work conditions) do not positively create satisfaction or motivation, whereas their absence causes dissatisfaction (Herzberg 1966). This dimension denotes basic needs that tend to be regarded as fundamental but may be taken for granted if they are currently met.

- *Resource* is defined by four needs: energy costs, avoiding wasting energy, the value or cost of the home, and concern for the environment. This dimension has a clear financial focus although 'waste' can also be seen from a non-financial perspective as something that is inherently wrong. It is interesting that concern for the environment fits in this dimension. This may be because the environment is seen as an external resource or because protecting the environment is seen as a consequence of the same actions that save money and avoid waste, rather than being a strong motivator in its own right.
- *Ease* is defined by four needs: doing what is easiest, keeping to everyday routines, doing what you have traditionally done, and doing what you think most people do. It represents convenience and simplicity, adopting (perceived) norms and other familiar behaviours. This need should not be seen as entirely negative: we would suffer cognitive overload if we had to think about what we are doing every time we do it and make new decisions about everything every day. Habits keep us sane (but not all habits are good).

These five dimensions and their linkages with the 21 needs are similar to the initial categorization developed from the qualitative research. That research suggested four broad categories of need (health and well-being, relational dynamics, agency and resources) of which eight sub-needs were found to have the most influence on daily, routine behavior (health and comfort, cost and waste, control and convenience, harmony and hospitality). The PCA of survey data identifies five dimensions of need

within the population as a whole. In essence, the 'health and well-being' category identified in the qualitative research was found to divide into two different dimensions – *Comfort* and *Hygiene*.

The qualitative research also found that households' hierarchy of needs shift over time – sometimes within a day. Short-term shifts can occur if, for example, caring for others is dominant only when others are present and health could become the dominant motive if somebody is feeling unwell at a particular time. In the long term, the needs of a household could vary with changes in the household composition or the age of the householders. This might suggest that the needs dimensions of the population will shift. While this cannot be tested directly by using the survey data, there are three reasons that it is unlikely.

- The survey asked for a general response rather than a response at a particular point in time, so short-term shifts should not be important.
- Respondents could choose any number of needs and so could identify any that were sufficiently important to them – they were not restricted to needs that are relevant at a particular time or in a particular context.
- While individual households might change over time, the survey provides data at a population (or sub-population) level rather than an individual level. Changes in individual households should therefore 'cancel out' so long as a sufficiently large number of households are included in the population or sub-population.

Reflecting on the third point, the needs should vary over a lifetime and that is part of their analytical value: it is possible to observe what is important to households that are at different life stages. This is observable and interpretable variation, rather than instability.

Characterizing households

The five dimensions of need offer a powerful and flexible means to characterize different households or population groups, as a guide for design and implementation of heating technology and strategies. In general, we found that the profile of needs varied with household characteristics and the heating system, rather than characteristics relating to the property. We found no marked differences between the needs profiles of those with different tenures, dwelling types or property ages. However, households identified by the presence of children, household size, education levels and household income varied quite markedly. Some examples are included below, demonstrating some logical relationships between group characteristics and needs profiles – relationships that promote greater confidence in the needs dimensions themselves.

Combining the needs dimensions with household dynamics reveals a consistency in survey responses and, in consequence, a new approach to thermal comfort. Figure 1 shows the needs profiles for the four household dynamics types. Scores on the dimensions have been standardized to emphasize variation among groups: a score of zero represents the average

for the population of British households as a whole; a higher score represents a greater relative emphasis on that dimension of need.

Figure 1 shows, first, that the two categories of ME household are rather different. Multi-person ME households (33% of households) are most distinct – primarily in the extent to which they prioritize *Ease* at the expense of *Other people*. This is not the case for JUST ME households (28%) who are rather more similar to the population average in their pattern of needs.

Key differences in the scores of the types of households exist on all of the five dimensions of need, with the exception of *Comfort*. The qualitative research suggests that comfort is the most fundamental need addressed by heating, so it is not surprising that it varies least across households that adopt different approaches to achieving comfort. YOU households (5%) are most distinct in the degree to which they prioritize *Other people* and, to a lesser extent, *Hygiene*, at the expense of *Ease* and *Resource*. The main difference with US households (34%) is that the latter are slightly less likely than average to prioritize *Hygiene* and more likely to prioritize *Resource*.

Figure 2 shows that relationships also exist between household composition and the five dimensions of need. Households

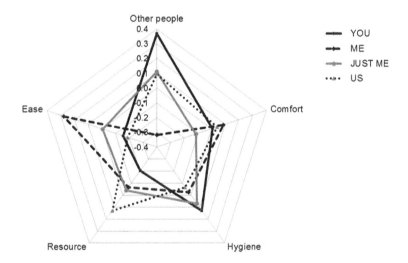

Figure 1. Needs profiles (relative factor scores) for four household dynamics types.

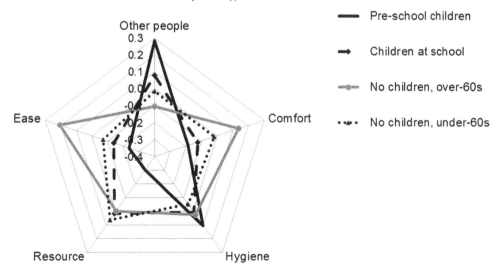

Figure 2. Needs profiles (relative factor scores) for four household compositions.

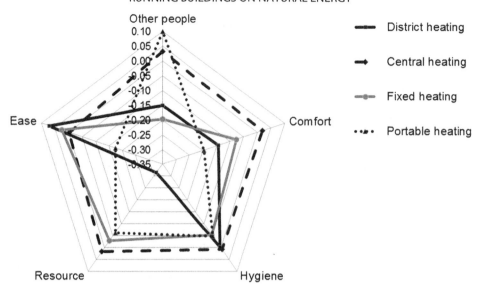

Figure 3. Needs profiles (relative factor scores) for four heating arrangements.

with children under school age (7% of the sample) stand out as having the greatest variation: this group prioritizes *Other people* and, to a lesser extent, *Hygiene*, at the expense of *Ease*, *Resource* and *Comfort*. Once children have reached school age (24%), *Hygiene* becomes less important and *Resource* more important, although both dimensions are close to average. Households containing all adults over 60 (38%) are rather different, prioritising *Ease* and *Comfort* at the expense of *Other people*. Households without children but with at least one adult under 60 (31%) are most similar to the population as a whole in their profile of needs.

Figure 3 presents scores across the five dimensions for four heating arrangements. Note that the scale is different from Figures 1 and 2 because the overall variation is less. District heating (2% of dwellings) refers to systems where the heat source is outside the dwelling, serving a group of dwellings (usually by hot water to radiators). Central heating (with or without some other form of heating) refers to the majority situation (84% of dwellings) where there is a central heat source for the individual dwelling. Dwellings without district or central heating relied on fixed individual heaters (e.g. gas, solid fuel or electric fires fixed in particular rooms) or portable heaters (e.g. a free-standing electric or bottled gas heater) – these dwellings represented 12% and 2% respectively of the total.

As might be expected, because households with central heating are in the majority, their profile of needs is similar to the average. Respondents with district heating placed the least emphasis on *Resource*, and low emphasis on both *Other people* and *Comfort*. This is perhaps because heat tends to be available all the time during the heating season but the household may have little control over the specification of this season and no access to thermostatic control. Households with portable heating also place little emphasis on *Comfort*, despite being the group that appears to have greatest difficulty keeping warm (44% of respondents with portable heating said they always keep warm enough by what they normally do on a typical winter day, compared with an average of 72%). This can be explained logically if (a) people who prioritize comfort do not use portable heating and/or (b) people with portable heating have learned

to limit comfort. Those using fixed individual heaters place little emphasis on *Other people*, whereas the opposite is true of those using portable heating.

Discussion

Understanding of the dynamics of energy behaviour offers insight into the various needs that can become motives for saving energy. Some practical application of this was described by Raw, Varnham, and Schooling (2010) and the new findings show how the motives/needs can be better categorized and quantified to support this application. The *Resource* motive (specifically saving money) has classically been used to promote energy-saving but motives relating to *Hygiene* and *Comfort* are more relevant to the indoor environment. Health and safety may be seen as fundamental motives for the use of energy although householders may not always conceive things in this way. For example, people use energy to cook food to eat and thus avoid starvation and prevent illness by killing pathogens. But the primary motive is more likely to be to avoid being hungry and/or to enjoy the food or the company around the meal table. Similarly, people may avoid illness by keeping warm but their primary motive may be comfort. So health and safety may be fundamental yet not necessarily prominent in conscious motive.

Some health and safety needs are met directly through energy use, such as keeping food fit for consumption and/or cooking it; keeping oneself and the home clean and dry; washing clothes or dishes; and using heat to relieve aches and pains. Meeting other needs depends on the indoor environment (e.g. warmth, unpolluted air and freedom from mould and decay). This is complicated by the fact that the indoor environment is affected by activities that have other main purposes (e.g. cooking or washing producing heat and moisture). Energy can be used for lighting to move safety around the home, to promote positive mood (often in combination with the thermal environment) or to enable certain tasks to be carried out (tasks that also have an optimal thermal environment). Health may also be affected by stress arising from not fulfilling other needs.

It is thermal comfort that has often dominated considerations related to energy use. However, *Comfort* is not just a striving after thermal neutrality – it includes: being relaxed and rested, free from worries or fears; a sense of physical and social 'cosiness' – the psychological comfort of a familiar setting that the person feels emotional attachment to; the physical comfort of resting after a day of activity, or the relief of the stress as the person relaxes; and the enjoyment that comes from pursuing individual or social interests when settled in a physically comfortable environment. This could entail, for example, using lighting, having a warm room, lighting a fire to create a cosy atmosphere, or relaxing in a warm bath.

On the specific subject of thermal comfort, it has been shown that a wide range of room temperatures can result in people reporting the same level of thermal comfort at home (Oseland and Raw 1991). This is not because the temperature has no effect but rather because individuals aim to achieve a temperature that is comfortable for them. In this sense, *comfort causes the temperature*, rather than vice versa. Therefore, the home should be designed to allow people to achieve comfort as they define it, which would entail a range of temperatures being achievable (where and when required in the home). Enhanced heating systems and controls can make it easier to achieve different temperatures efficiently in different rooms, or offer finer spatial control over radiant temperature and air movement.

But the use of space heating is not the only consideration. According to the adaptive model (Nicol, Humphreys, and Roaf 2012), the need for heating (or cooling) varies because the need for thermal comfort is met by means other than controlling the indoor environment. In a domestic context, our qualitative research (and Tweed et al. 2014) found that this variation is managed by individuals in the same space adopting different ways of keeping warm, for example: putting on extra layers of clothing and/or using extra bedding (on the bed or while seated during the day); using local heat from a radiant heater, hot water bottle, electric blanket or another person or pet; having hot food or drink; putting clothes, towels or bedding over a radiator to warm them before use; or having a hot bath or shower.

These alternative ways to keep warm are well known but too easily characterized as compromises for when the heating is inadequate. Not only can these alternatives save money (addressing the *Resource* motive), they can have merit in their own right because of the *Hygiene* and *Comfort* motives; the following are examples:

- Some individuals may enhance a sense of cosiness, and some find physical and psychological comfort, from behaviours such as wrapping up warm on a sofa to watch television, identifying this as self-indulgent and luxurious in a way that using heating may not be.
- Changing clothing can represent freedom from social expectations or from the belief that extra layers are uncomfortable or should not be necessary (clothing fabrics that appear light but are highly insulating may also be helpful).
- Adjusting clothing to the conditions can avoid disputes within the household about heating.
- Dressing warmly indoors in winter avoids the 'thermal shock' of entering or leaving a building where the temperature is

very different from outdoors, and makes it easier to maintain adequate ventilation during cooler weather.

Increasing metabolic rate (e.g. through exercise or physically active work) would also reduce the need for heating but motivations around this do not appear to have been studied, perhaps because it tends to be incidental rather than being used deliberately to keep warm. It is notable, however, that Gauthier and Shipworth (2015) report that the most frequent observed householder response to cold sensation was to change position or location. This might be related to some other action (e.g. getting some warm clothing or a hot drink, or moving to a warmer location) but there would in any case be a consequent increase in metabolic rate.

Hygiene and *Comfort* being motives for energy use, they can also be motives for actions that – as a by-product – also reduce energy use; the following are examples of this:

- Installation of insulation or efficient heating (with energy-efficient ventilation) can make a home more comfortable, double glazing also making the indoor environment quieter and more secure.
- Heat-recovery ventilation can improve indoor air quality and reduce problems with damp, mould and mites.
- Safety concerns can lead to replacing an old boiler (because of fears of fire, explosion or carbon monoxide poisoning); or limiting the temperature of radiators to avoid burns; or turning off appliances when not in use to reduce the risk of fire or lightning damage.
- Many people prefer the quality of daylight to that of electric lighting.
- Saving energy can improve well-being by making money available to meet other needs.

Conclusions

The identification of five underlying dimensions of need (in relation to heating the home and keeping warm) provides a means to look systematically at how energy advisors and heating system designers in particular can (a) motivate energy savings through addressing building users' needs and (b) maintain a good indoor environment, from the perspective of the users, while saving energy. This paper has focused on how two dimensions (*Comfort* and *Hygiene*) might be applied but the others (*Ease*, *Resource* and *Other people*) are also very relevant.

Thermal comfort specifically has to be seen in the context of other influences on heating strategies. This is perhaps obvious but there is now a greater degree of quantification about what such an assertion means. The need to be comfortable is the need that respondents most often identified as a *Big Factor* in determining how they use heating and keep warm. However, it was not the only consideration: concerns about energy cost or wastage are also major influences and each of 21 needs influenced at least 8% of respondents. Furthermore, whereas the specific need 'being comfortable' may relate primarily to thermal comfort, the *Comfort* dimension has broader connotations of being at ease, in control and free of concerns.

It is also important to understand the household dynamics that underlie decisions about heating and keeping warm:

heating is not a purely individual activity. The 'YOU–ME–US' typology is a simple way to characterize households, which can be overlaid on the dimensions of need. Comfort should be seen not simply as the thermal state of individuals but in terms of the actions of groups of individuals, in the context of conflicting needs and priorities.

The 'needs profiles' show that different groups emphasize different needs; one implication is that they may have different expectations or requirements for their means of keeping warm. This insight can support action at national or local level to reduce energy demand while also better meeting households' needs. The fact that household – rather than dwelling – characteristics have most influence on needs profiles suggests that selecting the best means to encourage consumers to save energy should be based more on the household than the dwelling, once the constraints of the dwelling have been taken into account. If surveys are conducted to determine the actual needs profiles of a household or defined population, this should be yet more effective than relying on household characteristics as a proxy.

Acknowledgements

The project as a whole was a partnership of PRP Architects and University College London, with NatCen Social Research as subcontractors. Under the guidance of the authors, the fieldwork was conducted by NatCen, with assistance from Seb Junemann of PRP Architects. Analysis was assisted by Matt Barnes of NatCen.

Disclosure statement

No potential conflict of interest was reported by the authors.

Funding

This research was commissioned and funded by the Energy Technologies Institute as a part of its Smart Systems and Heat Programme, Work Area 5: Consumer Response and Behaviour.

References

Gatersleben, B. 2001. "Sustainable Household Consumption and Quality of Life: The Acceptability of Sustainable Consumption Patterns and Consumer Policy Strategies." *International Journal of Environment & Pollution* 15 (2): 200–216. doi:10.1504/IJEP.2001.000596.

Gauthier, S., and D. Shipworth. 2015. "Behavioural Responses to Cold Thermal Discomfort." *Building Research & Information* 43 (3): 355–370. doi:10.1080/09613218.2015.1003277.

Herzberg, F. 1966. *Work and the Nature of Man.* Cleveland: World Publishing. OCLC 243610.

Lipson, M. 2015. *Smart Systems and Heat: Consumer Challenges for Low Carbon Heat.* ETI Insights Report. Birmingham, UK: Energy Technologies Institute.

Lorenzoni, I., S. Nicholson-Cole, and L. Whitmarsh. 2007. "Barriers Perceived to Engaging with Climate Change among the UK Public and Their Policy Implications." *Global Environmental Change* 17: 445–459. doi:10.1016/j.gloenvcha.2007.01.004.

Morgenstern, P. 2012. "Identifying Challenges to the Introduction of Heat Meters in Poorly-Performing District-Heated Apartment Blocks" (*MRes diss.*). The Energy Institute, University College London.

Nicol, F., M. Humphreys, and S. Roaf. 2012. *Adaptive Thermal Comfort: Principles and Practice.* Abingdon: Routledge.

Oseland, N. A., and G. J. Raw. 1991. "Thermal Comfort in Starter Homes in the UK." Proceedings of the annual conference of the environmental design research association, Oaxtepec, Mexico, 315–320.

Raw, G. J., and D. I. Ross. 2011. *Energy Demand Research Project: Final Analysis.* London: Ofgem. Accessed February 15 2017. http://www.ofgem.gov.uk/Pages/MoreInformation.aspx?docid = 21&refer = Sustainability/EDRP.

Raw, G. J., J. Varnham, and J. Schooling. 2010. *Focus on Behaviour Change – Reducing Energy Demand in Homes.* London: Department for Communities & Local Government. Accessed February 15 2017. http://www.communities.gov.uk/publications/planningandbuilding/energybehaviourchange.

Shipworth, M., S. K. Firth, M. I. Gentry, A. J. Wright, D. T. Shipworth, and K. J. Lomas. 2010. "Central Heating Thermostat Settings and Timing: Building Demographics." *Building Research & Information* 38 (1): 50–69. doi:10.1080/09613210903263007.

Tweed, C., D. Dixon, E. Hinton, and K. Bickerstaff. 2014. "Thermal Comfort Practices in the Home and Their Impact on Energy Consumption." *Architectural Engineering & Design Management* 10 (1–2): 1–24. doi:10.1080/17452007.2013.837243.

Upham, P., and C. Jones. 2012. "Don't Lock Me In: Public Opinion on the Prospective use of Waste Process Heat for District Heating." *Applied Energy* 89: 21–29. doi:10.1016/j.apenergy.2011.02.031.

Ventilation strategies for a warming world

Richard Aynsley and John J. Shiel ⓘD

ABSTRACT

World Surface temperatures have increased 20 times faster than they rose at the end of the last ice age, and they will continue to rise due to the volume of greenhouse gases already released into the troposphere. This paper explores the damage done to building occupants as widespread sealing is carried out for energy-efficient buildings in a warming world, with reduced ventilation to a point where toxic moulds grow in houses and offices and occupants develop sick-building-syndrome symptoms. Tools are also discussed to assist designers in accommodating rising indoor temperatures, and novel approaches are considered to improve ventilation, particularly providing guidance on the specific air gust frequency for increased cooling at low power.

Introduction

How much can ventilation strategies contribute to our future in a warming world?

There are two common philosophies related to designing buildings for dynamic external temperatures:

- Maintaining a narrow range of internal temperatures which requires a highly sealed envelope of low conductivity, mechanical heating and/or cooling, and a reliable power system despite severe storms, and
- A more passive approach where a wider range of temperatures are tolerated, and the temperatures are moderated with a low diffusivity building envelope (i.e. low conductivity/thermal mass) (Barrios et al. 2011).

World Surface temperatures have already increased 20 times faster than they rose at the end of the last ice age. They will also continue to rise due to the volume of greenhouse gases already released into the troposphere (Nuccitelli 2016).

Houses that are tightly sealed and highly insulated will be a liability (Lstiburek 2013) with a warming world if air-conditioning is relied on as the main mechanism for comfort and centralized power systems begin to fail due to:

- Severe weather events (Jacobs 2013a, 2013b; King et al. 2016),
- Dwindling fossil fuel resources (Mohr et al. 2015; WEC 2016), or
- Unforeseen circumstances (Kempton 2015).

So ventilation has a vital role to play in lowering temperatures that occupants effectively 'feel', particularly in heat waves and when indoor air quality deteriorates in humid areas such as the tropics and in coastal cities. More ventilation will be activated at night to cool buildings down after the heat of the day, and lifestyles are likely to change, with lighter clothing becoming more prominent at work (e.g. the Cool Biz campaign in Japan) and siestas implemented to minimize physical activity during the day, which are already popular in many countries. Greater use will be made of the cool stored a few metres below the surface, and in streams and groundwater sources. Ventilation is one of the few critical processes that can be performed passively by utilizing solar energy, wind, and gravity.

Recent global minimum ventilation problems

Prior to the 1970s, condensation in houses was rarely an issue. In those days, air leakage from a three bedroom house was between 900 and 1400 L/s (14 L/s per person), levels now considered indicate a *leaky* house. However, the 1970 Middle East 'oil embargo' brought with it increased energy costs and the need for energy conservation.

A widespread response to the oil embargo was an energy conservation programme for buildings. The installation of thermal insulation, together with severe cuts in ventilation, often to around 4 L/s per person or less, was introduced to reduce heating and cooling energy use. Sealing air leaks around windows and doors became a booming industry, and in large buildings, this reduced air leakage had consequences such as exacerbating 'Sick Building Syndrome' in office buildings due to poor indoor air quality (NHBC 2009; Wargocki 2013). Mould growth problems became commonplace, particularly in schools and houses due to indoor moisture build-up due to reduced ventilation. Many buildings with extreme mould cannot be reclaimed and are demolished (Brandon 2014). Building energy efficiency was gained, but at the cost of widespread poor indoor air quality.

It took decades after the oil embargo for building code authorities to accept the very poor state of indoor air quality and increase minimum ventilation rates, back to around

15 L/s per person (Janssen 1999). The resulting increased loss of heated or cooled indoor air by mechanical ventilation significantly increased building heating and cooling energy costs. Indoor air quality had become an issue with regard to indoor pollution from cigarette smoke and later, off-gassing of formaldehyde.

Surveys of houses suggested that air in a typical house is up to five times more polluted than surrounding outdoor air (Sherman 2004). With new houses being even more tightly sealed, expensive Energy Recovery Ventilation (ERV) systems were developed to recover heat or cool from exhausted air to pre-heat or precool incoming make-up air. More complex ERVs include moisture absorption from make-up air using desiccants.

Also, although earth sheltering to buildings can add thermal mass and assist with moderating internal diurnal temperature swings, it can also lead to greater exposure to harmful soil-borne gases such as radon and moist environments conducive to toxic mould growth.

Ventilation problems in New Zealand and Australia

Some countries have experienced building damage from interstitial condensation resulting from increases in thermal insulation and decreases in ventilation due to higher sealing requirements. In New Zealand, building damage resulting from energy efficiency flaws in their building code has been estimated to date at NZ$ 11.3 billion, for which the NZ government has been paying home owners compensation (Williamson 2009).

In addition, the Australia Building Codes Board guidance on condensation control in the latest Condensation Handbook (ABCB 2014) leads the housing construction industry towards wall construction commonly used in North America, which is mostly lightweight and heavily insulated (Aynsley 2012a, 2012b). However, in Australia there are three types of wall constructions (lightweight, medium weight, and heavyweight) that have roughly the same market share (DEWHA 2008).

For Australian homes with low ventilation rates, indoor air quality was also found to be compromised by pollutants generated within the house (Brown 2002; CSIRO and BoM 2010). This is concerning because buildings in Australia are being designed to match cool regions that use the German Energy Saving Regulation (EnEv) (Bambrook, Sproul, and Jacob 2011, 1706) or Passiv Haus standards (Ambrose, Syme, and DIIS 2015, 10), and very low air-tightness targets are being called for (Ren and Chen 2015, 79, 81) with little regard for ERV systems (Ambrose, Syme, and DIIS 2015, 10).

Indoor sourced pollutants that impact on the health of building occupants include carbon monoxide, carbon dioxide, radon, ozone, relative humidity less than 40% or over 60%, formaldehyde, bacteria, viruses, fungi, and dust mites and their droppings (Sterling, Arundel, and Sterling 1985). International indoor environmental quality and health studies have also shown that low ventilation rates in energy-efficient houses can have health and well-being consequences, and regulating bodies have set recommend minimum air exchange levels. Examples include:

- To prevent the risk of mould growth and poorer indoor air quality, a minimum air change rate of 0.5 air changes per hour at natural pressure (ACHnat) is recommended in the

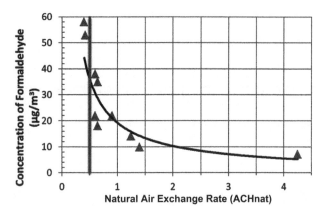

Figure 1. A survey of average concentrations of formaldehyde for US house air changes in natural air exchange rate (ACHnat). Source: Aynsley after Lstiburek (2013).

UK (NHBC 2009), which is approximately equivalent to 10 air changes per hour at 50 Pa (ACH50).
- Concentrations of formaldehyde significantly increased as the air change rate fell below 0.5 ACHnat (see Figure 1) when there was no mechanical assistance in a US survey by Offermann (2009). Nearly all houses in this survey failed a 9 µg/m³ 8-h Reference Exposure Level for formaldehyde (Offermann 2009, 5–6).
- Where 'Sick-Building-Syndrome' symptoms such as dryness, nasal problems and itching were reported across many international studies, leading to asthma, allergy and airway obstruction, the recommendation was that the ventilation rate should be greater than 0.4 ACHNat in existing homes (Wargocki 2013).

Formaldehyde standards

Formaldehyde standards include:

- The US, which has several standards, including an 8-h Reference Exposure Level for formaldehyde at 9 µg/m³ (Offermann 2009, 5–6).
- Japan, which has an exposure time of 1 h for a formaldehyde concentration of 100 µg/m³ and above in indoor air (Salthammer, Mentese, and Marutzky 2010).
- Australia regulations that limited the exposure to formaldehyde concentrations in indoor air in 1982 to less than 0.1 parts per million (120 µg/m³) (NHMRC 1982) and was revised in 2006 to less than 0.08 parts per million (100 µg/m³), similar to Japan's limit. However, in both cases no exposure time was nominated.

Insights into body heat exchange

Heat exchange between the human body and the surrounding environment is complex (Parsons 2003). The cooling capacity of airflow across exposed skin is yet to be fully utilized and will prove to be a key tool in accommodating rising global temperatures. The relative contribution of evaporative cooling by air movement across the exposed skin compared to other human body heat exchanges is shown by Chan (Chan 2008) in his diagram of a lightly clothed sedentary person. It shows that as the metabolism was maintained at a relatively constant level and operative temperatures increased:

- Combined radiative and convective heat exchange (losses) decreased up to an operative temperature of around 38°C and thereafter become heat gains.
- Evaporative cooling loss from respiration was relatively small and constant up to an operative temperature of around 28°C and thereafter heat loss increased with the onset of sweating.
- Body heat storage increased with operative temperature.

Tools for accommodating rising global temperatures

Thermal comfort and heat stress indices

The common thermal comfort indexes are Predicted Mean Vote, Effective Temperature (ET) with its variation Standard Effective Temperature (SET*) and Adaptive Thermal Comfort. The New Effective temperature variant, signified as ET*, is used close to the region of thermal neutrality, and cannot accommodate evaporative heat loss when operative temperatures exceed approximately 28°C. The Standard Effective Temperature signified as SET* includes the influence of skin wettedness, a key factor in the effectiveness of sweating for evaporative heat loss, as indicated by Chan above. Calculating SET* is complex; so computers are commonly used for this task. One readily available user-friendly computer program for this purpose is the ASHRAE 'Thermal Comfort Tool' (Shiel et al. 2017).

These thermal comfort indexes vary in ability to predict the cooling effect of airflow over exposed skin. A comparison of thermal-indexes-predicted thermal responses (Figure 2) indicates that the SET* index (the Pierce 2 Node Model) is closer to the empirical data (of KSU and Givoni et al. than to the Fanger Model).

Heat stress indexes are rational mathematical models used to predict the limiting metabolic rate due to heat stress in humans engaged in physical work in hot conditions. They are relevant to both inside buildings and outdoor locations. The common heat stress indexes are the ISO 7933 Standard, ACGIH index

referenced to the wet bulb globe temperature, USARIEM index and the ACP index. These heat stress indexes also vary in ability to predict the cooling effect of airflow over exposed skin on heat stress in subjects. A comparison of heat stress indexes' ability to predict metabolic rates in subjects under heat stress, and the influence of airflow (Brake and Bates 2002) indicates that the USARIEM index better models the influence of airflow to affect the limiting metabolic rate.

Weather monitoring and alerts

Just as bushfire alerts are now being sent by phone messaging, high forecasts of temperature and humidity conditions could be monitored and alerts sent to the regions affected. Strategies need to be put in place to evacuate those vulnerable especially if electricity failures occur as noted above.

Ventilation strategies

Natural ventilation

Air movement can be greatly helped with practical approaches to building design, including house orientation and its elevation, correct wing wall placement, extended roof overhangs, appropriate window types to catch breezes, dormer windows and wind towers, porous internal openings, correct spacing to neighbours, use of vegetation barriers and deflectors, buried cooling pipes, and installation of appropriate ceiling fans and their proper use (Aynsley 2014; Roaf 2012).

Much progress has also been achieved with the stack effect where there are a range of design options available to enhance air movement flows within a home, by adding roof-level openings and low-level intakes. Increasing interest is being shown in such methods by many designers in Australia and overseas for cooling-dominated climates, and especially in the light of global warming (Goad 2005, 25–26, 104; Parnell and Cole 1983, 59, 61, 79, 214; Roaf 2012, 113, 123, 292).

Mechanical ventilation

When installing and using residential ceiling fans:

- The fan motor should have a fine continuous speed control and large diameter blades, to save energy in summer, and avoid having to reverse the fan and reduce the heat loss through the ceiling in winter. This allows de-stratification of the warm air and can push it down near the occupants. Some manufacturers recommend reversing fans in winter since using the lowest of only three speeds will cause unwelcome drafts, but this is a very inefficient use of the fan and gives poor results with gusting.

Enhanced cooling effects of low-frequency gusts

Enhanced cooling effect of pulsing airflow was reported by Fanger and Pedersen (Olesen 1985). The increased cooling effect was found to be near the peak response frequency of thermal receptors beneath the skin, 0.47 Hz. Fanger and Pedersen (1977) suggested a method to include this effect in a building code by identifying an equivalent uniform velocity. A field study in China

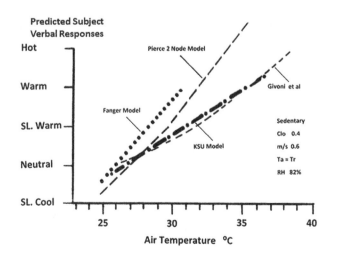

Figure 2. Comparison of predicted subject verbal responses from thermal comfort indexes from a common data set at 82% Relative Humidity (RH). Source: Aynsley, after Berglund (1978).

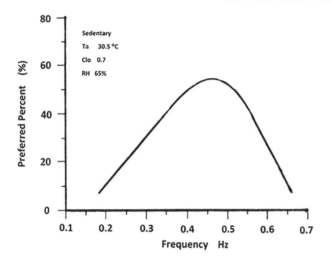

Figure 3. Distribution of preferred gust frequency at 30.5°C and 65% RH. Source: Aynsley after Xia et al. (2000).

Table 1. Uniform air speed required for equivalent cooling effect of 0.47 Hz air gusts.

Actual gust air speed at a gust frequency of 0.47 Hz (m/s)	Uniform air speed required for equivalent cooling effect (m/s)
0.10	0.24
0.15	0.3
0.22	0.36
0.29	0.45
0.40	0.6

Source: Aynsley, after Olesen (1985).

(Xia et al. 2000) illustrated in Figure 3 showed that subjects could dial up the mean wind speed and the gust frequency and that 95% of subjects were able to achieve thermal comfort at 30.5°C, RH 65% and Clo 0.7 within the gusting range up to 0.7 Hz. It also demonstrated that:

- the peak response frequency of human thermal receptors beneath the skin in hot humid conditions was the same as that for cold conditions, and
- the airflow with gust frequencies between 0.3 and 0.7 Hz is more effective at providing a cooling sensation than uniform airflow.

This low-frequency highly beneficial and low-power cooling effect is illustrated in Table 1 with the uniform air speeds that would be required for the equivalent cooling effects of a range of 0.47 Hz air gusts, and this has been exploited in:

- Car dashboard ventilation systems,
- Desk fans with a swing action, and
- With High-volume low-speed fans with a variable speed motor, for large spaces.

Comfortable buildings or comfortable occupants?

In a warmer future, the SET approach with occupants actively managing the building's environment can be used as an approach to have 'comfortable occupants' rather than 'comfortable buildings'. Instead of conditioning most of a building, occupants can make themselves comfortable with personal and building controls:

- In summer, e.g. with light clothing, small personal fans and even using the additional cooling effect of low-frequency gusting, ceiling fans on high speed, while at night in dwellings using ceiling fans on low speeds, and
- In winter, e.g. with heavier clothing, small radiant heaters under a desk, other heaters and fans on a very low speed to keep the heat near the ground and in dwellings using heavier blankets at night.

Better regulations

Occupants are spending more time at home by living longer and by working from home, or by having only part-time jobs, and so there should be:

- Better guidelines for condensation of typical Australian buildings,
- Indoor pollutant concentration and duration standards, especially for formaldehyde, and
- Minimum ventilation standards, including ones for homes with, and without, ERV systems.

Summary

Although heat stress indices and thermal comfort indices are similar, the critical difference is the criteria used in their application. Thermal comfort indices are typically used to ensure thermal comfort for around 80% of building occupants. Heat stress indices are typically used to determine a critical safe work time and rest time regimes for workers in hot environments in order to avoid heat stroke and other potentially lethal effects of heat. Neither of these indexes incorporates the confirmed benefit of airflow gust frequencies around 0.4 Hz probably because the initial research on the topic was focused on 0.4 Hz gust's detrimental effect in cold draughty conditions.

Other natural and mechanical ventilation enhancements outlined above can also achieve significant heating and cooling energy and greenhouse gas savings.

The response of building regulators in developed countries to save energy by requiring sealing of leaky houses and mandating higher levels of thermal insulation, without recognizing the risks of inadequate ventilation should be a lesson to us all. The only country that we know of that has seriously addressed this catastrophic bungle was New Zealand. Their government has paid out $11.3 billion dollars in compensation for damage to buildings. The damage to the health of occupants of these houses from indoor pollution remains unknown.

The construction of tightly sealed and highly insulated houses continues unabated, supported by an ever-increasing array of high-maintenance air-conditioning equipment. Builders in the residential market have shown little interest in availing themselves of more sustainable passive design techniques. In a similar manner, the air-conditioning industry has shown little interest in introducing increased air movement promoted by the

ASHRAE 55 Standard revisions to enhance summer comfort and reduce summer cooling loads. Passive ventilation seems to be a dying art in residential design. Ventilating strategies such as buried tubes for cooling make-up air used in the 1980s (Givoni 1994) come to mind.

Further research

Thermal comfort standards are continually updated with progress in relevant knowledge. Areas likely to influence thermal comfort standards for built environments are:

- Optimizing gust frequency of airflow devices to around 0.47 Hz based on its demonstrated benefits for summer cooling.
- Consider promoting the use of heat stress indexes in assessing thermal conditions in passively designed buildings to better acknowledge the risk of heat stroke and other heat-related effects, particularly on the elderly and infants.
- Upgrading ratings systems to consider the beneficial cooling effect of air movement, particularly at high humidity levels and particularly its contribution to the reduction of greenhouse gas emissions.

Conclusions

It is not clear that professionals in building design and construction have learned much from the widespread impact of poor ventilation in the quest for energy efficiency. More effort is needed to promote and demonstrate the benefits of adopting more passive and sustainable approaches such as earth cooling tubes to ventilation in building design and construction in our warming world.

ASHRAE has revised Standard 55 with new ASHRAE Thermal Comfort Tool software to promote the benefits of elevated airflow for cooling building occupants in summer. This tool needs to be adopted by building designers. Further increases in the cooling effect of airflow from specific gust frequencies around 0.5 Hz can be achieved by developing a standard for quantifying the presence of such gusting frequencies from airflow devices.

The quickest and most effective way to achieve genuinely low emissions buildings is to ensure that they are heated and/or cooled for as much of the day and year as possible with natural ventilation and much can be done to ensure this happens. The key step to take is to understand this simple reality. Once designers and decision-makers comprehend that, then rapid improvements in providing affordable comfort in more resilient buildings can be achieved by simply applying the idea using regulations, standards, guidance, tools, and through better design for natural ventilation.

Disclosure statement

No potential conflict of interest was reported by the authors.

ORCID

John J. Shiel ⓘ http://orcid.org/0000-0002-3233-6284

References

ABCB. 2014. *Condensation in Buildings.* Canberra: Australian Government and States and Territories of Australia. http://www.abcb.gov.au/ ~ /media/Files/Download%20Documents/Education%20and%20Training/Handbooks/Condensation%20in%20Buildings%202014%20second%20edition.pdf.

Ambrose, M., M. Syme, and DIIS. 2015. *House Energy Efficiency Inspections Project.* CSIRO, Australia: Commonwealth of Australia, Department of Industry, Innovation and Science & CSIRO. http://www.nathers.gov.au/sites/prod.nathers/files/publications/House%20Energy%20Efficiency%20Inspect%20Proj.pdf.

Arens, E., S. Turner, H. Zhang, and G. Paliaga. 2009. "Moving Air for Comfort." *ASHRAE Journal*: 18–28.

Aynsley, R. 2012a. "Condensation in Residential Buildings, Part 1 Review." *Ecolibrium* (October): 48–52.

Aynsley, R. 2012b. "How Much Do You Need to Know to Effectively Utilize Large Ceiling Fans?" *Architectural Science Review* 55 (1): 15–25. doi:10.1080/00038628.2011.641737.

Aynsley, R. 2014. "Global Warming Issues with Green Houses in Australia." Invited Keynote Talk presented at the Taishan Academic Forum, Jinan, China.

Bambrook, S. M., A. B. Sproul, and D. Jacob. 2011. "Design Optimisation for a Low Energy Home in Sydney." *Energy and Buildings* 43 (7): 1702–1711. doi:10.1016/j.enbuild.2011.03.013.

Barrios, G., G. Huelsz, R. Rechtman, and J. Rojas. 2011. "Wall/Roof Thermal Performance Differences Between Air-Conditioned and Non Air-Conditioned Rooms." *Energy and Buildings* 43 (1): 219–223. doi:10.1016/j.enbuild.2010.09.015.

Berglund, L. 1978. "Mathematical Models for Predicting the Thermal Comfort Response of Building Occupants." *ASHRAE Transactions* 84: 735–749.

Brake, R., and G. Bates. 2002. "A Valid Method for Comparing Rational and Empirical Heat Stress Indices." *Annals of Occupational Hygiene* 96 (2): 165–174.

Brandon, S. 2014. "Breaking the Mould." *Inside Housing*. http://www.insidehousing.co.uk/breaking-the-mould/7005664.article#.

Brown, S. K. 2002. "Volatile Organic Pollutants in New and Established Buildings in Melbourne, Australia." *Indoor Air* 12 (1): 55–63.

Chan, L. S. 2008. "Heat Exchange, Heat Exchange of Persons with the Environment." www.personal.cityu.edu.hk/ ~ bsapplec/heat.htm.

CSIRO, and BoM. 2010. *Indoor Air in Typical Australian Dwellings – Part 1.* Aspendale, VIC, Australia: Department of the Environment, Water, Heritage and the Arts. https://www.environment.gov.au/system/files/resources/87d5dedd-62c2-479c-a001-a667eae21f7c/files/indoor-air-project-dwellings.pdf.

DEWHA. 2008. *Energy Use in the Australian Residential Sector 1986–2020.* Australia: Department of the Environment, Water, Heritage and the Arts. http://industry.gov.au/Energy/EnergyEfficiency/StrategiesInitiatives/NationalConstructionCode/Documents/energyuseaustralianresidentialsector19862020part1.pdf.

Fanger, P. O., and C. J. K. Pedersen. 1977. "Discomfort Due to Air Velocities in Spaces." Proceedings of the Meeting of Commission B1, B2, E1 of the International Institute of Refrigeration, Paris, Vol. 4.

Givoni, B. 1994. *Passive Low Energy Cooling of Buildings* (263 pp.). New York: Van Nostrand Reinhold.

Goad, P. 2005. *Troppo Architects.* Singapore: Distributed by Tuttle.

Grimme, F., M. Laar, and C Moore. 2003. "Man & Climate – Are We Losing Our Climate Adaption?" Proceedings of RIO 3 – World Climate & Energy Event, December 1–5, Rio de Janeiro, 439–446.

Jacobs, M. 2013a. "13 of the Largest Power Outages in History – and What They Tell Us about the 2003 Northeast Blackout. Part 1." Union of Concerned Scientists – The Equation. http://blog.ucsusa.org/mike-jacobs/2003-northeast-blackout-and-13-of-the-largest-power-outages-in-history-199.

Jacobs, M. 2013b. "'Not A Good Day in the Neighborhood' – Electricity Grid Progress since the August 2003 Blackout. Part 2." Union of Concerned Scientists – The Equation. http://blog.ucsusa.org/mike-jacobs/electricity-grid-progress-since-the-august-2003-blackout-202?

Janssen, J. 1999. "The History of Ventilation and Temperature Control." *ASHRAE Journal* (September): 47–52. https://www.ashrae.org/File%20Library/docLib/Public/2003627102652_326.pdf.

Kempton, H. 2015. "Fault Found in Basslink Cable 100 km Offshore." *The Mercury*, December 22. http://www.themercury.com.au/news/tasmania/fault-found-in-basslink-cable-100km-offshore/news-story/38f74e0516c2745dcfa4bcc4808bde4a.

King, A., D. McConnell, H. Saddler, N. Ison, and R. Dargaville. 2016. "What Caused South Australia's State-Wide Blackout?" *The Conversation*. http://theconversation.com/what-caused-south-australias-state-wide-blackout-66268.

Lstiburek, J. 2013. "Deal with the Manure and Then Don't Suck." *ASHRAE Journal* (July). http://bookstore.ashrae.biz/journal/download.php?file = 2013July-040-046_building-sciences_lstiburek.pdf.

Mohr, S. H., J. Wang, G. Ellem, J. Ward, and D. Giurco. 2015. "Projection of World Fossil Fuels by Country." *Fuel* 141 (February): 120–135. doi:10.1016/j.fuel.2014.10.030.

NHBC. 2009. "Indoor Air Quality in Highly Energy Efficient Homes – A Review." Amersham UK: NHBC Foundation, prepared by BRE Building Technology Group. http://europeanparliamentgypsumforum.eu/wp-content/uploads/2012/04/Indoor-air-quality-in-highly-energy-efficient-homes.pdf.

NHMRC. 1982. *Urea Formaldehyde Foam Insulation*. 93rd Session. Canberra: National Health and Medical Research Council.

Nuccitelli, D. 2016. "Earth Is Warming 50x Faster than When It Comes Out of an Ice Age." *The Guardian*, February 25, sec. Environment. https://www.theguardian.com/environment/climate-consensus-97-per-cent/2016/feb/24/earth-is-warming-is-50x-faster-than-when-it-comes-out-of-an-ice-age.

Offermann, F. J. 2009. *Ventilation and Indoor Air Quality in New Homes*. Collaborative Report CEC-500-2009-085. PIER Energy-Related Environmental Research Program. San Francisco: California Air Resources Board and California Energy Commission. https://www.arb.ca.gov/research/apr/past/04-310.pdf.

Olesen, B. 1985. "Local Thermal Discomfort." *Bruel & Kjaer's Technical Review* (1), 23.

Parnell, M., and G. Cole. 1983. *Australian Solar Houses*. Leura, NSW: Second Back Row Press/Solar Scope.

Parsons, K. 2003. *Human Thermal Environments: The Effect of Hot, Moderate and Cold Environments on Human Health, Comfort and Performance*, 527. 2nd ed. New York: Taylor & Francis.

Ren, Z., and D. Chen. 2015. "Estimation of Air Infiltration for Australian Housing Energy Analysis." *Journal of Building Physics* 39 (1): 69–96. doi:10.1177/1744259114554970.

Roaf, S. 2012. *Ecohouse: A Design Guide*. 4th ed. Oxford, UK: Earthscan.

Salthammer, T., S. Mentese, and R. Marutzky. 2010. "Formaldehyde in the Indoor Environment." *Chemical Reviews* 110 (4): 2536–2572. doi:10.1021/cr800399g.

Sherman, M. 2004. "ASHRAE's New Residential Ventilation Standard." *ASHRAE Journal* (January): 149–156. https://www.ashrae.org/File%20Library/docLib/Public/20031231103644_266.pdf.

Shiel, J., R. Aynsley, B. Moghtaderi, and A. Page. 2017. "The Importance of Air Movement in Warmer Temperatures: A Novel SET* House Case Study." *Architectural Science Review*. http://dx.doi.org/10.1080/00038628.2017.1300763.

Shiel, J., B. Moghtaderi, R. Aynsley, and A. Page. 2014. "Reducing the Energy Consumption of Existing Residential Buildings for Climate Change and Scarce Resource Scenarios in 2050." In *Weather Matters for Energy*. New York: Springer. http://www.springer.com/us/book/9781461492207.

Sterling, E., A. Arundel, and T. Sterling. 1985. "Criteria for Human Exposure to Humidity in Occupied Buildings." *ASHRAE Transactions* 91 (Part 1): 611–622.

Wargocki, P. 2013. "The Effects of Ventilation in Homes on Health." *International Journal of Ventilation* 12 (2 September): 101–118.

WEC. 2016. *World Energy Trilemma 2016: Defining Measures to Accelerate the Energy Transition*. London, UK: World Energy Council. https://www.worldenergy.org/publications/2016/world-energy-trilemma-2016-defining-measures-to-accelerate-the-energy-transition/.

Williamson, M. 2009. "Government Announces Leaky Homes Package." http://www.beehive.govt.nz/release/government-announces-leaky-homes-package.

Xia, Y. Z., J. L. Niu, R. Y. Zhao, and J. Burnett. 2000. "Effects of Turbulent Air on Human Thermal Sensations in a Warm Isothermal Environment." *Indoor Air* 10 (4): 289–296.

Xia, Y., R. Zhao, and W Xu. 2000. "Human Thermal Sensation to Air Movement Frequency." Proceedings of Roomvent, International Conference, Reading, UK, 41–46.

Index

Printed and bound by CPI Group (UK) Ltd, Croydon, CR0 4YY

24/10/2024

01778287-0017